MW01280332

20

THE WHITE HOUSE

WASHINGTON

December 14, 2004

I send greetings to those celebrating the 100th anniversary of the Society of Automotive Engineers International. Congratulations on reaching this important milestone.

Safe, efficient, and reliable transportation systems are vital to our global economy. For a century, SAE International has helped advance mobility technology by providing industry standards, technical information, and resources for engineering professionals.

I applaud SAE International members for your hard work and commitment to excellence in your field. Your efforts help improve the transportation industry and serve the interests of people everywhere.

Laura joins me in sending our best wishes.

The SAE Story

One Hundred Years of Mobility

Robert C. Post

Tehabi Books developed, designed, and produced *The SAE Story: One Hundred Years of Mobility* and has conceived and produced many award-winning books that are recognized for their strong literary and visual content. Tehabi works with national and international publishers, corporations, institutions, and nonprofit groups to identify, develop, and implement comprehensive publishing programs. Tehabi Books is located in San Diego, California. www.tehabi.com

President and Publisher: Chris Capen
Senior Vice President: Sam Lewis
Vice President and Creative Director: Karla Olson
Director, Corporate Publishing: Chris Brimble

Senior Art Director: Josie Delker
Designer: Mark Santos

Editor: Katie Franco
Editorial Assistant: Emily Henning

Copy Editor: Mary Ann Short
Proofreader: Robin Witkin
Indexer: Broccoli Information Management

ISBN 0-7680-1489-1

First Edition
Printed by Regent Publishing Services Limited in China
10 9 8 7 6 5 4 3 2 1

Thanks to the SAE 100th Anniversary Committee, several dedicated SAE members, the board of directors, and SAE staff for their work in putting this book together.

SAE's one-hundred-year legacy of leadership in the mobility industry is visually depicted in this three-panel composition by John Glover. The artist's challenge was to create a two-dimensional celebration on canvas to honor the individuals and organizations that contributed to the success of SAE in its first century. This engaging triptych superimposes the common elements of time, people, education, technology, member services, and globalization over the three individually themed panels that include SAE's Automotive, Aerospace, and Heavy Commercial Groups.

Table of Contents

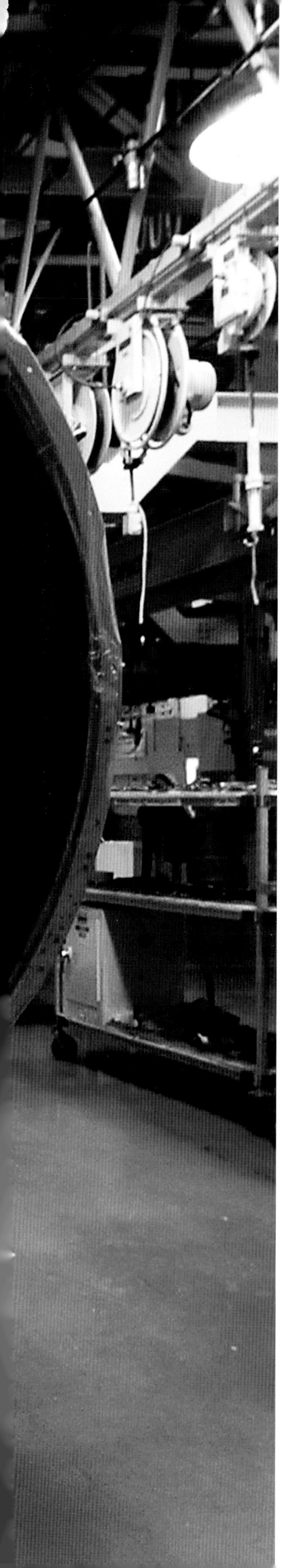

The Society of Automotive Engineers: Past, Present, and Future

THE FIRST ONE HUNDRED years of SAE's history have been filled with outstanding people, inventions, and new developments. Engineers desire to make something better through successive improvements in our technology, economy, and standard of living. Mobility engineers developed products and processes that led the shift from an agrarian to a highly industrialized economy. The measure of personal mobility has changed from five to ten miles per day by horse and buggy to continents per day by commercial airliner—or beyond with space vehicles.

SAE has contributed to these changes by being a forum for the interchange of ideas, technology, and knowledge and by developing standards that allow efficient industrial development of products and services. We are fortunate that many of the best engineering minds realized the value of interacting with their contemporaries under the auspices of SAE. Their interaction and exchange of nonproprietary information furthered development of the mobility knowledge base,

Opposite: The automotive industry has grown by leaps and bounds over the past century, and SAE continually adapts to meet the challenges of this ever-evolving industry.

A No-Till Drill plants soybeans in Ohio. SAE has been meeting the needs of the heavy-duty industry for almost one hundred years.

Previous spread: A General Electric aircraft technician inspects the engine of a Boeing 767.

products, and services, and have contributed greatly to progress and the standard of living we enjoy in the world today.

Today, SAE has grown to be a global organization serving the worldwide needs of our constituents: members, companies, academia, and government. The different mobility sectors linked together under the term *automotive* have grown to include everything from small to heavy-duty engines, off-road to space vehicles, and fuels and lubes to electronics. This diversity of members and technologies is one of SAE's greatest strengths and has played an important role in the growth of all these mobility areas.

SAE's voluntary consensus standards process with approximately fifteen thousand participants has produced over seventy-five hundred standards used daily in the design, manufacturing, marketing, service, support, and recycling of mobility products. Our more than eighty-five thousand members and others are engaged in lifelong learning through education offered in the AWIM program, Collegiate Design Series competitions, conferences, technical meetings, and professional development series.

In the next one hundred years the only certainty is that change will continue at an accelerated pace. Our successors will develop technologies that far exceed our best current technology. SAE must continue to adapt and change the products and services we provide as the needs of our constituents change. SAE has a bright future because our two fundamental core competencies, lifelong learning and standards, will be even more highly valued in the years to come. Please join me in reflecting on the past one hundred years and working together to make the next one hundred years even better for SAE and society.

DUANE TIEDE

2004 SAE PRESIDENT

RAMSTAD HVASS
WEST 1/4
11th-Grant Sts 1/2
5th Ave S 3/4
DLI
SHOULDER USE PERMITTED WHEN METERED

From Dusk to Dawn: Tradition, Transition, and Transformation

AS THE SUN SETS on the first one hundred years of SAE, we reflect on a proud and rich tradition of contribution to the transportation industry. Engineering accomplishments in the last century have been the enabler of progress, expanding our personal horizons on land, in the sky, and beyond. At the start of the century, four U.S. farmers could feed about ten people. By the end, with the help of engineering innovation, a single farmer could feed more than one hundred. In the early 1900s, average Americans traveled about twelve hundred miles in their lifetime, mostly on foot, and mostly within their own town. One hundred years later, the typical American adult travels more than twelve thousand miles by automobile alone in just one year. From early test rockets to sophisticated satellites, the human exploration into space is perhaps the most amazing engineering feat of the twentieth century. The development of spacecraft has thrilled the world, expanded our knowledge base, and improved our capabilities.

Previous spread: Traffic whizzes in and out of Minneapolis, Minnesota.

SAE has come a long way since mobility pioneers Andrew Riker, Henry Ford, E. T. Birdsall, and others met in New York to discuss the need for a society dedicated to automobile engineers. From early luminaries like Orville Wright, Amelia Earhart, Charles Kettering, and Howard Coffin, through today's engineers working in many industries, SAE has assisted, influenced, guided, followed, and molded mobility engineers all over the world. Ever since 1916, when the Society of Automobile Engineers merged with the Society of Tractor Engineers and the American Society of Aeronautic Engineers, SAE has played an important role in the growth of these mobility areas.

What a legacy the men and women of SAE—some of the most extraordinary inventors of the twentieth century—have left for us. They created engineering marvels resulting in new products, new technologies, and new applications that have done nothing less than change the course of society. This kind of quantum leap in engineering evolution comes from the discussion and sharing of science and technology—the very cornerstone of SAE's mission. Even as business and communication models change, the exchange of ideas opens the door for great things. We will continue to bring together experts from around the world to share knowledge and present their latest developments, continue to publish technical papers, gather the engineering community's best minds for collaboration and insight, establish the best practices and standards for all industry and government, and develop ongoing education forums for all engineers to improve their skills through classes and technical sessions. These are the traditional areas of SAE contributions that create a lasting foundation for our future.

As with any great milestone, one hundred years is an occasion to celebrate. We should take the time to reflect on our accomplishments and, perhaps more important, apply the collective learning and knowledge to the next hundred years. As the sun rises on this new century, we are welcomed with a new, global horizon with continuously expanding curiosities and conundrums. With each new daybreak comes evolving social, environmental, and business trends that present not only challenges, but also opportunities.

With North American and European transportation growth reaching a plateau, sustainability of the industry turns to developing transportation in countries like China, India, and others. China's automotive demand is approaching five million vehicles a year, and joint ventures and partnerships with technologically based Asia–Pacific corporations are the order of the day for product development and design. Increasing demand for vehicles begets increasing demand for highway construction and development of infrastructure. All of these developments open the door for SAE involvement, sharing and consulting on the engineering challenges for vehicle design and engineering, as well as off-highway transportation to support the massive construction efforts. China is not alone in its construction explosion: with industrial growth and expansion across the globe, SAE's growing membership from more than one hundred countries worldwide can provide the leadership and institutional memory to new members and their companies in all areas of mobility engineering.

The automotive industry, with almost sixty million vehicles produced around the world each year, is under continuous pressure from consumers and government regulators to advance developments in passive and active safety systems, raise emissions standards, produce leading-edge product design, and explore new propulsion systems. This pressure to innovate is shared by both manufacturers and suppliers. As chief engineer emeritus, General Motors, I now have the opportunity to view engineering from the supplier side of the business, and I can testify to the increased engineering responsibility of automotive suppliers, who now provide two-thirds of all innovation in vehicles.

In the aerospace industry, we see the development of highly efficient propulsion systems as a dominating force as well. With the industry evolving to global businesses and global communications, the restrictions of time and distance are shrinking rapidly and propulsion demands will increase. As a recognized leader in developing standards for the global aerospace industry, SAE will continue its role to support and lead our members' efforts.

SAE is educating tomorrow's leaders, who are today's elementary school students learning the thrill of engineering through our groundbreaking A World in Motion programs. They are high school students learning more advanced skills in more complex versions of A World in Motion II programs. They are college students putting all their skills and knowledge to test in our many collegiate design competitions. They are freshly minted professional engineers attending the SAE Congress and other events to learn the newest technological advances. They are engineers of all ages and skill levels using SAE books, papers, and standards; attending continuing education classes; presenting papers at an SAE conference, and accessing leading information on SAE's Web-based products to stay at the top of their fields.

As it was at the time SAE was first created, we are again in a time of transformation, on the cusp of engineering breakthroughs that will once again expand our personal horizons, shrinking the size of the world through transportation.

Ultimately, SAE must transform itself during the next century as businesses, governments, and communication methods are transformed with new technologies and demands. The future has never shone brighter for SAE and its ability to contribute and serve society. Please join with me in our year of celebration of the past, and the future.

Ted Robertson

J. E. "TED" ROBERTSON

2005 SAE PRESIDENT

1

Making History

The automobile is European by birth, American by adoption. The internal-combustion engine, upon which most automobile development has been based, is unmistakably of European origin, and both the idea and the technique of applying it to a highway vehicle were worked out in Europe. On the other hand, the transformation of the automobile from a luxury for the few to a convenience for the many was definitely an American achievement, and from it flowed economic consequences of almost incalculable magnitude.

JOHN B. RAE, *The American Automobile* (1965)

An 1885 Scientific American *article about Carl Benz's three-wheeler, above, inspired Charles and Frank Duryea, left, to build the first American auto powered by internal combustion. Here they pose in 1895 near their shop in Springfield, Massachusetts.*

"The car that put America on wheels." In 1921 Henry Ford poses with a new Model T, one of about one million produced that year and fifteen million between 1908 and 1927.

Etienne Lenoir's stationary engines burned a mixture of air and illuminating gas.

WE THE UNDERSIGNED desiring to form a corporation pursuant to the provisions of the Membership Corporations Law . . . do hereby certify and state:

FIRST: *The particular objects for which the corporation is to be formed are: to promote the Arts and Sciences connected with Engineering and the Mechanical construction of Automobile vehicles; to hold meetings for the reading and discussion of papers relating to the construction and improvement of automobiles and to promote social intercourse among members of said Society; to publish and distribute papers and literature relating to the construction of automobiles and to maintain an Engineering Library; to grant certificates of membership . . . .*

SECOND: *The name of the proposed corporation is* THE SOCIETY OF AUTOMOBILE ENGINEERS:

SIGNED

Henry Hess
Edward T. Birdsall
Alexander Churchward
A. H. Whiting
Horace M. Swetland

Emile Levassor is seen here with his hand on the tiller and his back to Madame Levassor, who is seated next to Levassor's associate Rene Panhard. This vehicle predates the 1891 Panhard et Levassor, with its front-mounted engine turning the rear axle through a drive-shaft, which became conventional engineering practice.

The Origins of the Auto

Steam-powered conveyances first performed for excited crowds more than two hundred years ago, before the turn of the nineteenth century. By the 1840s experimental locomotives with battery-powered electric motors had made an appearance. Internal combustion, the technology that would trump them both, was a latecomer—Etienne Lenoir's two-stroke appearing in 1860, Nicholas Otto's four-cycle gasoline-burning version in 1876. The direct precursors of the motor vehicle were created independently in Germany a decade later by two men whose names still resonate in the world of automobility: Gottlieb Daimler, who had been an engineer in Otto's firm, and Carl Benz, a manufacturer of stationary engines. Daimler's was a 1.5-horsepower two-wheeler with a high-speed (600 rpm) engine and hot-tube ignition, Benz's a 0.8-horsepower three-wheeler with a spark ignition and differential rear axle.

For a time, technological leadership passed to France, where advancements came at the hands of Emile Levassor, Arnold Peugeot, and others. Back in Germany, Benz debuted an improved four-wheeled vehicle in 1893, and in the United States that same year, Charles and Frank Duryea finished building the first practical automobile—they used that very term—and drove it through the streets of Springfield, Massachusetts. By 1896 the Duryeas could build one of their automobiles in thirty days. Within six years Ransom E. Olds was producing and selling more than two hundred a month, twenty-five hundred all told in 1902. Ten years after that, Henry Ford's factory in Detroit could turn out thousands in one day, and in ten more years, half the cars on American roads would be Model T Fords.

Rudyard Kipling dismissed the automobile as a "petro-piddling monster." Maybe that's the way it looked to an English man of letters, but in the United States there was nothing piddling about the auto. The industry grew with incredible speed: ranked only as the nation's 150th largest by the 1900 census, it had made first place by the 1920s. Growth might have been even faster but for a controversial patent that covered the mechanical features of a road vehicle powered by an internal-combustion engine.

The patent holder was George B. Selden, a patent attorney in Rochester, New York. The existence of the Selden patent led to the creation of an organization called the Association of Licensed

During the protracted litigation over the Selden patent, Hugo V. Gibson drives a vehicle allegedly conforming to the patent's provisions along 56th Street in New York City. The man at left, partially obscured, is Charles E. Duryea, a witness for Henry Ford; those walking alongside in suits are Edward M. Bentley, a witness for Selden; and Harry T. Cointon, a representative of the ALAM.

Automobile Manufacturers (ALAM), which demanded a royalty of 1.25 percent on the list price of every auto with an internal-combustion engine. Only Henry Ford and his outspoken associate James Couzens held out. "Selden can take his patent and go to hell with it," roared Couzens. Not until 1910, after a trial that lasted years and filled dozens of volumes with testimony, was the royalty rendered ineffective.

By then another organization was starting to flourish, one with more high-minded purposes than exacting (Couzens would have said extorting) royalties from automakers. This was the Society of Automobile Engineers (SAE). Peter Heldt, editor of *The Horseless Age,* had proposed the creation of a professional society of automobile engineers in 1902:

> *Now that there is a noticeable tendency for automobile manufacturers to follow certain accepted lines of construction, technical questions constantly arise which require for their solution the cooperation of the technical men connected with the industry. These questions could best be dealt with by a technical society. . . . Meetings could be held at specified intervals, at different places, and papers read and discussed on subjects relating to the branch of engineering which the society represents.*

For most of the nineteenth century, the U.S. Patent Office required that a model be submitted along with a patent application. Selden's model was the handiwork of William Gomm, a Rochester machinist whose experiments had proven to Selden that the optimum fuel for an internal combustion engine was gasoline.

The Selden Patent

The names of the earliest American autos are richly evocative: Stanley, Duryea, Winton, and Olds in the 1890s; Buick, Pierce-Arrow, Packard, Cadillac, Ford, Reo, Pontiac (neé Oakland), Hupp, and Hudson in the first decade of the twentieth century. Nearly all the manufacturers of these autos accepted an obligation to pay a percentage of their profits to an organization called the Association of Licensed Automobile Manufacturers (ALAM), which held rights to a patent that had been secured by George B. Selden.

Selden's patent for a "road engine" covered a combination of elements: a motor using liquid hydrocarbon fuel, a transmission, and a steering mechanism. Selden's motor was a version of the Brayton Ready Motor, a stationary unit whose charge of burning fuel exerted constant pressure on the piston, much in the way a gas turbine operates. Even though the capabilities of Selden's motor were limited, they were sufficient to win him his patent, which he kept pending for sixteen years, until November 5, 1895, by which time he had modified it to cover compression engines such as Otto's and Daimler's. It was numbered 564,160.

But Selden was of a sort who are legion in the annals of technology, a better dreamer than doer. He frightened potential investors with wild predictions that motor vehicles would soon outnumber horses on city streets and he could not secure financing for a prototype. In 1899 he sold his patent to a Wall Street syndicate for ten thousand dollars plus 20 percent of potential royalties. The syndicate then began infringement proceedings against manufacturers, including Alexander Winton, the nation's major producer of internal-combustion motor vehicles. With a view toward negotiated settlements, control of the patent was turned over to the ALAM, which issued licenses that obliged every automaker to pay a 1.25 percent royalty on every sale.

Seeing it as a way of warding off fly-by-nights, manufacturers generally went along with the ALAM, even though they did not accept the patent's validity. An exception was Henry Ford, who was ambivalent about patents and allowed free use of his patented devices. Soon Ford enlisted allies such as Ransom E. Olds in fighting the ALAM through another organization, the American Motor Car Manufacturers Association. After eight years of court battles, the Selden patent was upheld. The victory was moot, however, because the patent was shortly due to expire. The more lasting effect of the controversy was to transform the popular perception of Ford into a champion of the little guy, and indeed his status as a folk hero has its roots in the Selden patent case.

With his share of license fees collected by the ALAM, George Selden moved into manufacturing in 1906, but completed only two Selden Buggys before going bankrupt. The date painted on the side was especially annoying to Henry Ford because nothing, not even the engine, was any older than 1904.

Opposite: *Wilbur and Orville Wright are seated on the steps of their home in June 1909, not long after their demonstration to the War Department had proved the capabilities of their flying machine. Wilbur, older by four years, would die in 1912 at age forty five; Orville would live until 1948.*

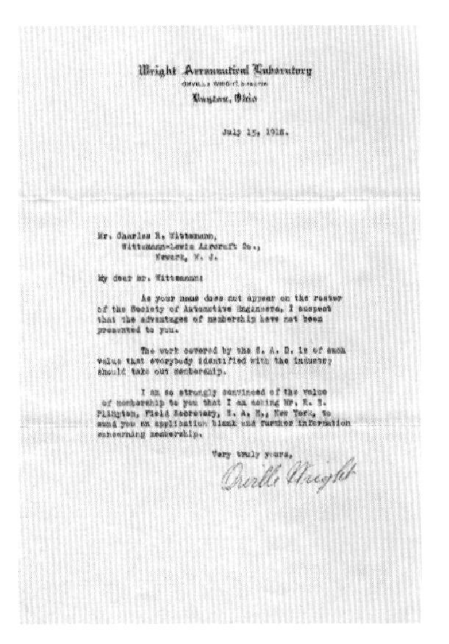

Wright Aeronautical Laboratory

Dayton, Ohio

July 15, 1918.

Mr. Charles R. Wittemann,
Wittemann-Lewis Aircraft Co.,
Newark, N. J.

My dear Mr. Wittemann:

As your name does not appear on the roster of the Society of Automotive Engineers, I suspect that the advantages of membership have not been presented to you.

The work covered by the S. A. E. is of such value that everybody identified with the industry should take out membership.

I am so strongly convinced of the value of membership to you that I am asking Mr. [illegible], Field Secretary, S. A. E., New York, to send you an application blank and further information concerning membership.

Very truly yours,

Orville Wright

Orville Wright sent this letter to Charles Witteman on July 15, 1918.

The Horseless Age

Surely there were plenty of technical questions in need of addressing: Chain drive or differentials? Planetary transmissions or sliding gears? Sawdust for quieting noisy gearboxes? Disc or cone clutches? Four-wheel brakes? Shock absorbers? Make-and-break or hot-tube ignition? Steam, electric, or internal-combustion power? This last question generated a great deal of debate. And perhaps most important, how were nomenclature and technical specifications going to be standardized? Would automobile manufacturers and engineers willingly share knowledge? In Europe, automakers were supplying certificates attesting to the rated horsepower and torque curve of their vehicles. There was nothing like that in the United States.

The Society of Automobile Engineers

It did not take long for Heldt's idea about the cooperation of technical men to begin materializing. Essential energy was provided by Edward Tracy Birdsall, a consulting engineer. Birdsall was also a charter member of the American Institute of Electrical Engineers who later worked for Henry Ford and still later for the Wright Aeronautical Company in Dayton.

Birdsall's link with Wright Aeronautical is interesting on two counts. First, he initially suggested the idea of what would become the Society of Automobile Engineers in a letter to prominent engineers dated just eighteen days before Orville and Wilbur Wright made history in Kitty Hawk, North Carolina. Second, it would not be long before the shared concerns of automobile and aeronautical engineers became so apparent that they would merge under the banner of SAE. One of the treasures in the SAE archives is a note on Wright Aeronautical Laboratory letterhead, addressed to Charles Witteman of the Witteman-Lewis Aircraft Company and signed by Orville Wright. It begins:

> *My dear Mr. Witteman:*
>
> *As your name does not appear on the roster of the Society of Automotive Engineers, I suspect that the advantages of membership have not been presented to you.*
>
> *The work covered by the S.A.E. is of such value that everybody identified with the industry should take out membership.*

Some testimonial! But this is getting ahead of the story (as is the change in name to Automotive). In January 1904, during the New York Automobile Show, Birdsall invited "fellow

Diverse Origins

Most of the first automakers were expert machinists, intimate with shop culture, especially shops that produced other sorts of conveyances—bicycles, wagons, carriages. Some were successful entrepreneurs: Albert A. Pope was the foremost manufacturer of bicycles in the United States, and the Studebaker brothers were the world's leading producer of horse-drawn conveyances. The White brothers turned to the manufacture of steam-powered autos from sewing machines. Ransom E. Olds had been producing stationary gas engines. Walter P. Chrysler was involved in the railroad locomotive industry. David D. Buick made plumbing fixtures. George Pierce of Pierce-Arrow went from making birdcages to bicycle spokes to complete bicycles to automobiles. It is remarkable that the automobile industry was successful despite its varied beginnings.

Above right: Roy Chapin, then a gear-filer for Ransom E. Olds, poses in the curved-dash Olds he drove from Detroit to New York, 820 miles, for the opening of the 1902 National Automobile Show. This model was manufactured between 1901 and 1906, retailing for $650; in 1902 it accounted for 28 percent of total U.S. production.

Right: Louis Renault, whose background was quite similar to Henry Ford's, is seen here in his workshop in France in 1898. Renault was the first French automobile manufacturer to set up an assembly line, then an integrated manufacturing operation, and by 1908 he was the French industry leader.

Previous spread: Kill Devil Hills, North Carolina, December 17, 1903, 10:35 AM. Orville Wright, on the lower wing, is airborne at last; older brother Wilbur stands aside, and their tools lie scattered in the sand.

conspirators" to meet in the offices of the Automobile Club of America at 58th Street and Fifth Avenue to organize a society that welcomed all engineers concerned with the design and construction of automobiles.

Henry Ford poses in 1902 with his first "quadricycle," completed in 1896, the same year that Hiram Maxim, Ransom E. Olds, Alexander Winton, and other Americans also built their first autos. Essentially a motorized horse buggy, Ford's and the others were a long way behind French state-of-the-art conveyances.

The men who met with Birdsall included Andrew L. Riker, who had designed a gasoline engine for the Locomobile, and was vice president of that company in Bridgeport, Connecticut; Hiram P. Maxim, who had been chief motor-vehicle engineer of the Pope Manufacturing Company in Hartford, a giant among bicycle manufacturers that had diversified into both electric- and gasoline-powered autos; Horace Swetland, publisher of a trade journal called *The Automobile* (later renamed *Automotive Industries*); H. F. Donaldson, an associate of Swetland's; Henri Chatain, who was with General Electric in Schenectady; Henry Vanderbeek, who was with Henry Timken's new roller-bearing company in Canton, Ohio; and Allen Whiting, the New York City manufacturer of an auto that bore his name, the Whiting. Also invited was another man who had affixed his name to an automobile, Henry Ford, but Ford, ever the maverick, did not show up. Those who did attend appointed a committee to choose a name for the organization and formulate a constitution and bylaws.

In January 1905, the thirty founding members of the new Society of Automobile Engineers again convened in New York and elected officers. Riker became president; Ford, first vice president; and John Wilkenson, designer of the air-cooled engine used in the Franklin, second vice president. Birdsall, whom the others had quickly spotted as a workhorse, was named secretary-treasurer. Attendees also agreed to solicit members by word of mouth and in Swetland's journal, *The Automobile*.

When the members next met in 1906, they had to confess that the membership campaign had not been particularly successful: SAE still had only fifty-two members. The thirty-two who attended came from nine different states, with the largest contingents from New York and Connecticut. The best news was Birdsall's report of a cash balance of $343.66, and with that, SAE elected to launch a publication. Volume 1, no. 1, of *S.A.E. Transactions* appeared in print later that

By 1907, when this photo was taken, the National Automobile Show at Madison Square Garden in New York had become a major industry event.

year with three papers: one on materials by Thomas Fay, who would succeed Riker as SAE's president; one on ball bearings by Henry Hess, who would succeed Fay as the third president; and one on carburetor design by Birdsall himself.

By 1907 when the membership had increased to one hundred, the reading of four papers used the time allotted for the business meeting and voting for new officers. Those currently in office were carried over for another year.

The Society continued to grow in 1908 and 1909, but slowly. Birdsall later remarked that his main worry was that someone would start a rival organization. When someone did, Birdsall dismissed it as a group of high-grade chauffeurs, and the organization faded. Still, many automobile engineers already belonged to the mechanical branch of the American Society of Mechanical Engineers (ASME) and could not see much point in paying ten-dollar dues to a struggling organization that as yet offered few benefits. Birdsall later penned a reminiscence that is wonderfully evocative of a renowned society's modest beginnings. As secretary-treasurer, he wrote:

> *I was necessarily the goat. They shoved all the work on me. I had to get the members, and then collect the money from them; and, believe me, it was no easy work in those days to collect* $10 *a year when we were not giving the members anything in return. In those days we did not even have a badge. As Secretary I would get an application for membership. As the Chairman of the Membership Committee I would pass on the member. As Secretary, I would send out notice of his election. As Treasurer, I would send him a bill for his dues. If the Membership Committee thought that the member proposed was not eligible, as Treasurer I would say to myself, "That's all right, but we need the money." And the Treasurer would prevail on the Chairman of the Membership Committee to elect the man to membership.*

Bicycle-Making Heritage

Many people who got SAE up and running had been in business making bicycles—Alexander Winton, the Duryea brothers, and William Knudson, for example—and one should not forget the Wright brothers, who brought to airplane design a deep familiarity with bicycle technology. In *The Automobile Age*, James J. Flink writes:

> *Apart from its impact on road improvement in the United States, no preceding technological innovation—not even the internal combustion engine—was as important to the development of the automobile as the bicycle. Key elements of automotive technology that were first employed in the bicycle included steel-tube framing, ball bearings, chain drive, and differential gearing. An innovation of particular note is the pneumatic bicycle tire. . . . The bicycle industry also developed techniques of quantity production using special machine tools, sheet metal stamping, and electric resistance welding that would become essential elements of the volume production of motor vehicles. . . . The greatest contribution of the bicycle, however, was that it created an enormous demand for individualized, long-distance transportation that could only be satisfied by the mass-production of motor vehicles.*

Among key aspects of automobile engineering that were borrowed from the bicycle, perhaps the most important was the so-called balloon tire. Although he was not the first to secure a patent on inflatable tires, this invention is usually attributed to John B. Dunlop, a native of Dreghorn, Ayshire, who worked as a veterinarian. Dunlop is the luxuriously bearded man on the right.

Opposite: Elwood P. Haynes and Elmer and Edgar Apperson finished building their automobile on July 3, 1893, and joined the Duryea brothers as the only Americans who had so far built an automobile. It is seen here in 1894 on the streets of Chicago.

The Society finally incorporated in 1909, at which time it also debuted an SAE logo, but sustained growth began only after the breakup of the ALAM. One of the most important consequences of the Selden patent controversy had been to alert people in the automobile industry to the danger of protracted patent litigation becoming a financial drain. The result was a system of mutual cross-licensing, supervised initially by the National Automobile Chamber of Commerce, the successor to the ALAM and predecessor of the Automobile Manufacturers Association. Of the major manufacturers, only Packard and Ford steered clear of this agreement, although Ford was a nominal participant.

There was another more portentous consequence of the Selden patent controversy. SAE's takeover of the defunct ALAM's technical section was the beginning of the Society's standardization program, the activity that put it on the map to stay.

Andrew L. Riker

In a hundred years, SAE has had only one president who served more than one year, its first, A. L. Riker. After graduating from Columbia University law school in 1888 at the age of twenty, Riker began making electric motors in Brooklyn. Ten years later, he organized the Riker Electric Vehicle Company. After leaving that firm, he developed a rather advanced internal-combustion engine and interested the Locomobile Company in manufacturing it in both two- and four-cylinder versions for autos aimed at a high-end market. During the three years Riker was president of SAE, he was also vice president of Locomobile, and this was when owning a Locomobile meant owning the very best.

Like Coker Clarkson, another giant in the early years of SAE, Riker died in 1930, just after returning home to Connecticut from French Lick, Indiana, where he had joined in celebrating the Society's twenty-fifth anniversary.

Standardization

> *Standardization and interchangeability of parts will have the effect of giving us a higher grade of motorcar at a lower price, but this is dependent to considerable degree upon the production of one model in great numbers and elimination of extensive annual changes in design that necessitate the making of costly jigs, gauges, and special machinery.*
>
> *Scientific American*, January 16, 1909

Standardization had roots in the manufacture of muskets with interchangeable parts at the turn of the nineteenth century, later in the navy with items like pulley blocks, and in the railroad industry's standard designs for such components as couplers. There had been efforts by other engineering societies to standardize screw threads and pipe diameters and to establish ratings for the capacity of electrical equipment. While interindustry technical standards were still a rarity at the turn of the

Opposite: Howard Coffin, president of SAE in 1910, poses in the late 1920s with U.S. president Calvin Coolidge at the entrance to Coffin's Georgia retreat on Sapelo Island, much of which Coffin owned.

Auto racer Louis Chevrolet poses in his prototype passenger car in 1911. The first Chevrolet came off the line in 1912.

twentieth century, automobile manufacture seemed like fertile ground for them. Automaking was typified by small firms dependent on other suppliers for parts and components and auto repair by efforts to improve procedures for servicing and repair that entailed standardization of lubricants, wheel rims, and the like.

Beginning in 1905, the ALAM Mechanical Branch sought to establish standards for spark plugs and screw threads, along with specifications for alloy steels. These efforts were dropped in 1909, but standardization had been a primary concern within SAE ever since its founding four years earlier, and external factors quickly contributed to the creation of a successful program. A depression in 1907 and 1908 precipitated the first merger in the automobile industry, the formation of General Motors (GM) out of four other companies: Buick, Cadillac, Oakland (later renamed Pontiac), and Oldsmobile. By 1910 GM controlled more than 20 percent of the nation's total automobile production. The Ford Motor Company had another 10 percent. Small independent manufacturers (more accurately termed *assemblers*) faced growing difficulties in obtaining credit and capital, and in 1910 eighteen of them went bankrupt, a shocking number because the average had been only one per year from 1903 to 1909.

Second-tier companies, the most in distress, badly needed a standardization program, and it was these same companies whose engineers and executives dominated the membership rolls of SAE. Howard Coffin, who became SAE's fourth president in 1910, understood the problem both as an engineer and as vice president of the Hudson Motor Car Company. His partners were the Detroit department-store magnate Joseph Hudson (after whom the car was named) and Roy Chapin, famed for driving a curved-dash Olds from Detroit to the 1901 New York Automobile Show in only ten days. In 1912 the Hudson Motor Car Company produced five thousand vehicles, not insubstantial but far fewer than Ford's seventy-eight thousand. Hudson needed standardization.

"No other organization," Coffin told SAE's members, "has ever had such an opportunity . . . to wield its influence for the direct and immediate good of that business which has called it into being." The Standards Committee established by Coffin was headed by Henry Souther, a graduate of the Massachusetts Institute of Technology (MIT) and SAE founder who had been the foremost engineering specialist at the ALAM. Like Coffin, he saw the establishment of standards as crucial to the survival of the smaller manufacturers. As Coffin put it in a 1910 article in *S.A.E. Transactions*,

Opposite: Women machinists are seen here circa 1918 on the shop floor of the Minneapolis Steel and Machinery Company, which claimed to pay women at the same rate as it paid men performing the same work.

the lack of standardized components "was responsible for nine-tenths of the production troubles and most of the needless expense entailed in the manufacture of motorcars." SAE published its first standard in 1912.

While chairing the SAE Standards Committee from 1910 to 1915, Souther created sixteen standards divisions, from "Aluminum and Copper Alloys" to "Wood Wheel Dimensionings and Fastenings for Solid Tires." One major goal was to reduce the vast number of minute variations. For example, the industry had been using more than three hundred types of lock washers; this number was cut by almost 90 percent. SAE standards committees also reduced sixteen hundred sizes of seamless steel tubing (with as many as eighty different sizes in a single vehicle) to 221. Dimensional standards for rims, spark plugs, carburetor flanges, generator mounts, and many other components also were aimed at minimizing variety. There were efforts to gain general acceptance for the H-shape pattern of gearshift-lever movement. SAE standard specifications were set for lubricants, ferrous alloys, and, of course, screw threads, nuts and bolts, and splines.

While there was never any chance of enforcing coercive measures, SAE's standardization program was welcomed by much of the industry. By 1916, shortly after SAE opened a Detroit office, more than 90 percent of the respondents to a Society questionnaire indicated that they adhered to the standards for nuts, bolts, lock washers, wheels and rims, and spark plugs. A 1921 article in *Automotive Industries* magazine estimated that the SAE standards were responsible for savings of $750 million, or 15 percent of the retail value of all automobiles sold.

Still, even as the SAE program had resulted in 224 sets of industry standards by 1921, neither intercompany standardization nor cross-licensing of patents was able to keep all the small producers competitive with Ford and GM. These firms depended less on outside suppliers, and when they did buy components on the open market, they bought in large quantities that gave them the

Coker Clarkson

Coker Clarkson was SAE secretary and general manager for twenty years, 1910–1930. He died suddenly while still holding that office at the age of sixty. His SAE memorialist sketched some of his attributes:

> *Deeply studious, with a prodigious capacity for silent work . . . inflexible in principle; filled with an inner sense of justice which all near to him recognized; incapable of affectation or self-italics; saturated with the sense of impersonal labor and purpose which alone makes a true executive.*
>
> *Clarkson drew to him the admiration and affection of the varied types of the huge new industry to whose imperative problems he addressed himself incessantly during the most vital period of our times. Upon his field of endeavor he left an impression which will be more accurately valued as the titanic processes to American industry unfold themselves to the succeeding generation.*

To which his dear friend and early champion Howard Coffin added:

> *Words are weak things with which to pay tribute to Coker Clarkson. He has been for 20 years one of the most constructive figures in American industry. His remarkable qualities of leadership and his quiet efficiency, as well as his singular and unselfish devotion to his life's great work, have made for him a lasting place in the hearts of all of us whose privilege it has been to know him. He has our admiration for his accomplishments, our respect for his ideals and our love for him as a friend. The structure he has built will live on in the useful fulfillment of his vision, and his memory will remain an inspiration to us always.*

The "chute" at Ford's factory in Highland Park, Michigan, where bodies were dropped temporarily onto chassis, after which they were hauled to the loading dock and chassis, bod-

Mass Production

Before the automobile, most conveyances had been produced with handicraft methods, one at a time. But the auto was the ultimate product of a new system dating to the latter part of the eighteenth century, when manufacture entered a phase based on the utilization of new sources of power, new kinds of machinery, and new ways of organizing the workforce, including the mechanized production of devices from components that were identical and thus interchangeable.

During the nineteenth century, especially in the United States, new production methods were applied to a growing range of consumer products and were inevitable with automobiles. In 1908 three autos were sent to England, dismantled, and the parts mixed together. When reassembled, all three started right up and ran perfectly. It happened that these were Cadillacs, but that same year, Henry Ford introduced the Model T, the vehicle that linked the concept of mass production to Ford's name so securely that in other countries it was known as "Fordism."

Ford remarked that the way to tap a mass market was "to make one automobile like another automobile, to make them all alike, to make them come from the factory just alike—just like one pin is like another pin when it comes from the pin factory." In 1913 his brain trust came up with the idea of a moving assembly line, with each worker "stationed at a specific point along the line and responsible for only one small step" in assembling Ford's Model T; often a worker who put in a bolt did not put on the nut. The numbers tell the story: in 1908, when Model Ts were $850, Ford sold 5,986 of them; in 1916, when they were $350, Ford sold 600,000. By 1920 half the autos in the world were Model Ts.

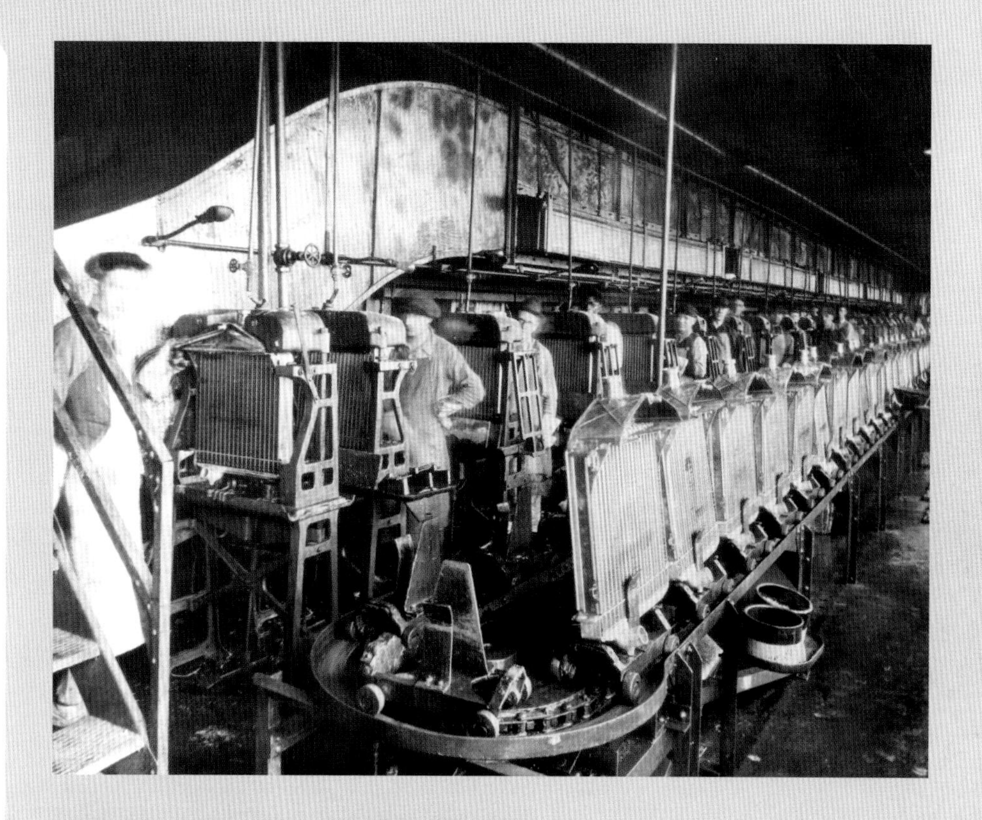

Workers assemble radiators, above, and flywheels, left, in factory assembly lines. These mass-produced autos changed America and eventually the world.

E
45

E 45
E 46
E 47
E 46
E 47

Opposite: Women taking the wheel: In contrast to the photo at left, here the driver commands an upscale right-hand-drive automobile.

The automobile comes to rural America: a scene in a North Dakota town, circa 1910.

Previous spread: A stern-faced foreman eyes a team of the most highly skilled Highland Park workers, the men on the Model T Upholstery Line.

leverage to negotiate prices. As the historian John B. Rae put it in *The American Automobile,* "the big companies . . . could achieve the necessary efficiency in production by internal standardization, which gave them greater flexibility in controlling their own operations than conforming to industry-wide specifications would have permitted."

As the number of smaller manufacturers dwindled during the 1920s, so too did the interest in intraindustry standards. Only later did the largest firms become interested in standardization, as management realized that standard specs for tires and petroleum products would help sell motor vehicles by rendering them more user friendly. By the 1930s the standards that received most attention were those concerned with basic engineering practices and efficient operation, whereas the sort of detail that had been so important to small producers—standards for door hinges, generator brushes, and the like—were mostly gone from the Society's annual enumeration of standards, the *S.A.E. Handbook.* Nevertheless, the SAE program already represented a major achievement. The Society played a significant role in the creation of the American Standards Association. The public was fully accustomed to seeing "SAE Standard" on bottles and cans of motor oil (petroleum refiners had begun to embrace SAE specifications for viscosity as early as 1911). And SAE specifications were absolutely essential to the aircraft industry.

Membership

Before 1910 only halfhearted efforts were made to increase SAE membership rolls: $250 was collected at the 1908 meeting in Cleveland for "propaganda," and a clearinghouse for job information was established as an aid to recruitment as well. But the slow rate of growth mirrored slow growth in the industry. Production in the United States totaled twenty-four thousand autos in 1905 and only sixty-three thousand by 1908. There was a notable jump in 1909—the result, among

The Membership Debate

William Van Dervoort, who headed the Moline Auto Company in Illinois, was SAE president in 1915. In his presidential address, he advocated "a most careful scrutiny of all applicants" and suggested that membership drives were not even necessary. Then he went on to say: "I look forward to the day when the standards of this institution will be such, and the honor of becoming a member so great, as to raise the requirements of admission to the point where only those with the best training and experience can hope to attain full membership."

Charles Manly was the first engineer from the aviation industry to be elected president, four years after Van Dervoort. In his presidential address, Manly urged that the basis for membership not be "how great [the] distress a man has passed through, either in the way of recognized course of instruction, or professional association or apprenticeship." An "aristocracy or inner circle," he continued, one promoted by overly stringent qualifications for membership, was likely to result in most members having "one foot in the grave." After much discussion, SAE followed Manly's advice, and members are welcome in all stages of their careers.

Above: On December 14, 1903, Smithsonian Institution Secretary Samuel P. Langley, right, posed with his intrepid pilot and mechanic Charles Manly before an attempt to get Langley's Aerodrome airborne by means of a catapult mounted on the roof of a houseboat on the Potomac River.

other things, of the Model T coming into production (though Henry Ford did not remain active in SAE, he gained lasting renown in Society annals for a paper titled "Simplicity"). But the first big jump in SAE membership took place between 1910 and 1912, when the organization went from six hundred to thirteen hundred members. And that was due specifically to the standardization program that Howard Coffin and Henry Souther launched in 1910 with the hiring of an Iowan named Coker Clarkson as SAE's salaried secretary and general manager.

A scenic overlook along the Laurel Canyon road in Hollywood, California.

Clarkson was exceptionally able—he had studied philosophy at Harvard with William James, had a law degree, had been secretary of the mechanical branch of the ALAM, and had edited the annual *Handbook of Gasoline Automobiles.* Even so, there was a great deal of opposition to his proposed salary of five thousand dollars, which was more than the Society's annual income from dues and initiation fees; Coffin, along with Riker and Swetland, had to guarantee his salary personally. But the hiring of Clarkson and the Society's renewed commitment to standards provided a major impetus for a membership campaign—standards now gave engineers a substantial inducement for joining, as evidenced by the 200 percent growth in just two years. Coffin, Souther, and Clarkson began to envision an SAE that was at least the equal of the so-called founder societies.

It was not all smooth sailing. In addition to questions the standards program raised about engineering principles and practices, major disarray in internal policy had to be sorted out. When Clarkson first surveyed the whole range of SAE activities, he confirmed Coffin's apprehension that the records were confused, the finances inadequate, and most disturbing, the membership dissatisfied. The "Brief History of the Society of Automotive Engineers" from 1946 reports on a vexed 1910 meeting of the council:

> *President Coffin proposed to SAE council that if the Society were to properly serve automobile engineers, to establish automobile engineering as a profession, and to function without suffering from the effects of commercialism so disruptive of other organizations, new policies and new methods were mandatory. A stormy session of SAE Council ensued. Questions of constitutionality and of financial procedure were raised. Impeachment of the new president was suggested.*

Mass Motorization

By the 1890s auto manufacture was well established in Europe: Carl Benz produced more than eleven hundred autos before 1898, and in France the firm of De Dion-Bouton turned out more than fifteen thousand between 1895 and 1901. At least fifty-nine British firms had begun making autos by 1901. It took the Americans a while to catch up—production in the United States did not surpass French production until 1904—but then the Americans roared past the rest of the world. In 1913 the United States had one auto for every seventy-seven people; in Great Britain the ratio was 1:165. By 1927 the ratio was 1:5 in the United States, 1:44 in Great Britain, and 1:196 in Germany.

After William C. Durant lost control of General Motors, he formed Durant Motors, Inc., with plants in Flint and Lansing, Michigan, and in Long Island City, New York. Seen here in the early 1920s is the so-called Drive Away outside the Lansing factory. Durant is the man facing the camera.

Times Square in New York City, looking uptown from 43rd Street in the late 1920s. Automobiles vie for space with pedestrians and electric streetcars, the latter destined to disappear within a few years in order to make room for more motor vehicles.

In 1908, at one of his Dearborn, Michigan, farms, a youthful Henry Ford takes a ride on one of his Ford-powered "automobile plows."

Growth in membership led to qualifications becoming a hotly debated issue: Should membership be limited to engineers with professional credentials (estimated at two thousand in the United States), or should it be constituted on a broader basis? Should "persons of prominence" in the auto industry who were not trained engineers be admitted to full membership? A strong difference of opinion emerged.

The SAE constitution specified "persons connected with the arts and sciences relating to the engineering and mechanical construction of automobile vehicles" as qualified for membership, and a Society faction even thought in terms of restricting membership to men who had designed a complete automobile. Such a person was "a very rare article," said a representative of the opposing faction, Henry Hess. An executive with the New York Power Department, former president of the Society, and proponent of a broad-based membership, Hess remarked that many potential members were inhibited from applying because they were unsure of their qualifications and perhaps "too modest."

The upshot of this dispute was a revision to the constitution, in 1911, to make it more in keeping with the wishes of liberalizers like Hess. A regular member could be engaged in the design or manufacture of parts or accessories and not necessarily entire motor vehicles. An associate member no longer had to qualify as a regular member or be "competent to cooperate" with engineers. Rather, an associate could be anyone "in a responsible commercial or financial capacity," a purchasing agent, say, who might have sound advice about specifications, or even a race-car driver—an unusual occupation, but growing—who might be an authority on engine design. Even though associate members could not vote or hold office, many who were not engineers began joining the Society. They were not engineers in the sense of being practitioners of actual mechanical construction but people with titles such as sales engineer and a great many who identified themselves as occupying some executive capacity.

Federal Aid for Roads

Though the national government had appropriated funds for turnpikes in the early days of the republic, no such financing had been made available since the 1840s. In 1907 only 7 percent of the roads in the United States had any surfacing at all, and that was usually just gravel. Then in 1916 Congress voted to allocate seventy-five million dollars over a five-year period for the improvement of post roads. This was followed in 1921 by legislation that provided matching grants to states, to promote an interconnected national system of highways, and more than ten thousand miles were completed the next year alone. Seldom was a new road not swarming with traffic soon after it was finished.

A 1919 coast-to-coast convoy aimed at influencing legislation in favor of completing the Lincoln Highway, a feat that was accomplished in 1925. Left: Some of the convoy's seventy-two vehicles make a rest stop near Big Springs, Nebraska. Above: A truck sporting an Army Corps of Engineers insignia on its canvas top and towing a watercraft negotiates a rugged detour near Sterling, Illinois.

Opposite: The 1904 World's Fair, known as the Louisiana Purchase Exposition, featured a New York–to–St. Louis automobile parade. Autos are seen here arriving at the exposition against the backdrop of another fairly recent invention, the Ferris Wheel.

After Charles and Frank Duryea dissolved their partnership in 1898, Charles moved from Springfield to Reading, Pennsylvania, and established the Duryea Power Company; one of its horseless carriages was the first ever to be equipped with pneumatic tires.

Inevitably, some of the old-timers raised the specter of "industry domination," especially with the creation of another membership class called affiliate that was specifically for manufacturing firms. For fifty dollars (double the cost of a regular membership), a firm could become an affiliate and designate up to six representatives whose dues were ten dollars each. Concern about the professional interests of engineers and those of manufacturers not necessarily coinciding was nothing new. As early as 1907, Thomas Fay had read a paper in which he kept referring to "Mr. Commercial Man," someone who might compel engineers to do things that did not "look right." Later, in 1911, during a debate about relaxing the requirements for becoming an associate, one member expressed similar concerns about "elements" becoming dominant in the Society who would not be interested first and foremost in best-practice engineering.

Henry Hess remarked in response that company managers could control the organization from without, simply because they provided so many of the employment opportunities for engineers. Concerns about industry domination were not borne out, in any event, at least not in the sense that "Mr. Commercial Man" proved overpowering. During the 1920s, when the Society began its monthly *Journal*, it regularly printed a list of new members and their occupations. Practically all were engineers. Those in favor of admitting nonengineers as associates did have a powerful set of arguments, however: that standardization work required substantial funds for testing, that only by broadening the membership base could this funding be assured, and that the continual elaboration of new standards clearly was the wave of SAE's future.

By 1910 when SAE emerged as a full-fledged professional society under the leadership of Coffin, Souther, and Clarkson, the motor vehicle had become almost commonplace in American life. Said Charles E. Duryea: "The novelty of owning an automobile has largely worn off. The neighbors have one of their own. The whole family has become so accustomed to auto riding that some members generally prefer to ride alone or remain behind while others go."

SAE Becomes Automotive

The whole family was not accustomed to riding in airplanes, however, and Theodore Roosevelt gave Americans a scare in 1910 when he took a ride in a Wright biplane. Even so, in 1911 a suggestion was aired at SAE's annual meeting for establishing a branch for aeronautical engineers, in emulation of

Above: A proud farmer shows off his rugged new Farmall tractor and reaper. The decision to add agricultural and other engineers proved wise for SAE: by the time this photo was taken in 1920, membership had topped five thousand.

Previous spread: At the St. Louis Aerial Meet on October 11, 1910, Theodore Roosevelt settled into a Wright biplane next to pilot Arch Hoxsey. The aim was to dispel the apprehension that air travel was dangerous, but Roosevelt's flight was nonetheless worrisome; among the hundred fliers who died in mishaps the next year was Mr. Hoxsey.

one of the Society's German counterparts, the Verein Deutscher Ingenieure (VDI). During a debate over admitting engineers from other realms, Coffin warned that it would be disastrous for automobile engineers to presume that they were in a class by themselves. At the 1916 meeting, the council voted to admit, without initiation fee, members of not only the American Society of Aeronautical Engineers but also other members of other societies who had been knocking on SAE's door: the Society of Tractor Engineers, the National Association of Engine and Boat Manufacturers, the National Gas Engine Association, and the American Society of Agricultural Engineers.

In 1917 Elmer Sperry—an inventor whose eminence at the time rivaled Edison's—coined the word *automotive* to cover any kind of self-propelled vehicle and to replace *automobile* in SAE's name. This represented a final rebuff to proponents of automobile exclusivity, and SAE gained one thousand members during 1917 alone. In 1920 membership reached five thousand, including the Society's first woman member, Nellie M. Scott. By that time standardization work extended to all industries involved in making vehicles powered by internal-combustion engines or making parts for such vehicles.

The larger firms, however, were the last to get deeply involved. Before World War I the two industry giants, Ford and GM, seemed to get along without not only SAE and its standards but also without trained professional engineers. Henry Ford did not much like engineers, and William C. Durant of GM didn't bother with them. Durant himself had only a smattering of technical knowledge, and his waste of money on such oddball enthusiasms as friction drive was a major factor in his losing control of GM. James J. Storrow, who handled GM's reorganization after Durant's departure, immediately put on a drive to hire graduate engineers but also moved men without formal training into key administrative positions, notably Walter P. Chrysler. His pioneering GM Research Department did not pay much attention to SAE standards at first; of the roughly three hundred men serving on SAE standardization committees at any given time, only a handful were from GM Research.

Lawrence Sperry, Elmer Sperry's son, poses with his invention: detachable landing gear combined with landing skids. Early aviation pioneers helped the American Society of Aeronautical Engineers merge with SAE, opening the door for future generations of aviation specialists.

In some quarters there was a fear of what was termed *overstandardization*, and *S.A.E. Transactions* had to offer assurances that the concept of "one model in great numbers" was not meant to limit inventiveness. Still, one could find occasional descriptions of the "ideal car," with an optimum displacement, wheelbase, and so forth. These proved worrisome to those who admired novelty, but needlessly so, because relatively few components had been standardized by the 1920s, compared with a half century later when the standards book would run to more than a thousand pages.

A War of Transportation

In January 1916 SAE hosted what it termed a preparedness banquet featuring speakers Josephus Daniels, secretary of the U.S. Navy, and General Leonard Wood, chief of staff of the U.S. Army. Daniels asked for the Society's assistance in developing standardized powerplants for airplanes and submarines; Wood stressed the need for rugged, agile trucks—likewise standardized—and urged automakers to prepare to convert their factories to war production.

For SAE World War I provided a marvelous opportunity to demonstrate the practical value of its work; Secretary of War Newton D. Baker described the fighting as "a war of transportation."

The Liberty Truck SAE helped produce went through rigorous testing before it was used in World War I. Here the truck is being tested on its ability to cross over railroad tracks.

In 1914 a new class of membership called *departmental* had been created for representatives of government departments and bureaus at all levels, and this marked the beginning of efforts to secure federal appropriations for standardization work. By the time the United States entered the war as an associated power in April 1917, there was already an SAE Washington office and lines of communication were in place so that cooperative efforts could be launched.

At SAE's 1917 meeting nearly every paper addressed military technology and several were delivered by representatives of government agencies. Six of the eleven past presidents of the Society went to work for the U.S. government: Hudson's Howard Coffin chaired the Advisory Committee of the Council of National Defense as well as the Aircraft Production Board and the Industrial Preparedness Committee of the Naval Consulting Board. Locomobile's Andrew Riker also served on the Naval Consulting Board. Howard Marmon of the Nordyke-Marmon Company worked with the Aviation Section of the Navy Department. William Van Dervoort, SAE's president in 1915, served on the Council of National Defense and the Aviation Section of the Signal Corps, as did Henry Souther, now an independent consultant. Timken's H. W. Alden was also involved in the development of military technology, as were four other men later to become president of SAE.

Cooperation between government and SAE proved quite effective. SAE's new Aeronautical Engines Division worked with the U.S. Navy Department and the Signal Corps to develop the Liberty airplane engine; another division worked with the National Bureau of Standards on octane ratings and viscosities. Perhaps most significant were the efforts of the Truck Standards Division and the War Department to develop standard vehicles called Liberty Trucks for the Quartermaster Corps: although standardization of truck components had already gone a long way before the war because most manufacturing was done on a small scale, large advances were made during the war.

Obligations to the Public

The development of the Liberty Engine and the Liberty Truck garnered a great deal of favorable publicity for SAE, and after the war the Society took understandable pride in a level of governmental recognition not attained by any other engineering society. There was a distinct change in

The Great War

The small-scale manufacturers that dominated SAE played a major role in preparing for U.S. involvement in the war in Europe. Henry Ford, who kept his own counsel on many matters, had pacifist leanings and at first opposed the preparedness effort. So did William Durant, who in June 1917 refused to convert his Cadillac factory to the production of Liberty aircraft engines, designed by Packard engineers. Although Durant later changed his mind, by then Henry Leland had resigned from the company to form the Lincoln Motor Car Company, where he produced thousands of Liberty Engines, as did Marmon and several other firms, even Ford. GM and Dodge concentrated on munitions, whereas Ford production was as varied as engines, armor plate, tanks, caissons, submarine chasers, and torpedo tubes. All this was in addition to the hundreds of thousands of trucks and ambulances that rolled out of hundreds of U.S. factories.

Above: This poster shows how heavily soldiers at the front and civilians behind the lines depended on trucks for provisions during World War I.

Left: Together with the airplane, the armored, self-propelled tank freed modern warfare from protracted stalemates. Although the tank's first major success, at Cambrai in November 1917, was achieved by three hundred British tanks, tanks were also manufactured by American automakers, like the one pictured here by the Ford Motor Company.

Trucking Takes Off

Shortly after the turn of the twentieth century, studies showed that a five-ton truck was more expensive to operate than a wagon with a three-horse team, given the same load, but that the truck's greater speed resulted in less than half the cost that of the wagon and team. The lesson was not lost on any organization responsible for large numbers of wagons and teams—such as the U.S. Army. Even before the United States entered the war in 1917, more than forty thousand U.S.-made trucks and ambulances had been delivered to the Allies in Europe. By the end of the war, the U.S. armed forces were operating three times that many.

In 1918 Reo, White, Locomobile, Packard, Peerless, Autocar, and nearly 350 other U.S. manufacturers delivered two hundred thousand trucks to the army, and the majority were exported to the war zone. A member of the British War Cabinet remarked that "the war could not have been won if it had not been for the great fleets of motor trucks." The Allied cause "floated to victory on a wave of oil."

Short hauls by trucks were not uncommon before World War I, but the war marked the turning point for over-the-road trucking, when U.S. railroads proved inadequate for transporting material to East Coast military embarkation centers without inordinate delays in marshaling yards. By the winter of 1917, convoys of trucks were headed eastward from midwestern factories and certain roads were declared National Military Truck Routes.

August Fruehauf introduced tractors and semitrailers, Fageol Truck and Coach Company introduced aluminum bodies. Kenworth—a firm founded by two Seattle maritime clerks, H. W. Kent and E. K. Worthington—introduced specialized log trucks. Chain-driven Mack AB and AC models so impressed British soldiers that they were dubbed "bulldogs." After the war, army surplus gave domestic trucking a major boost, and truckers were instrumental in securing passage of the Federal Highway Act of 1921.

By then more than a half million trucks traveled U.S. roads—many of them built to rugged army specifications that SAE had been instrumental in developing—and in the mid-1920s long-haul trucking began to take off as the number of trucks registered in the United States topped two million. During the Great Depression, intercity truck traffic more than doubled.

Motorized vehicles used in farming and for hauling over-the-road trailers were both known as "tractors." Here, in 1907, a farm tractor tows a wagon stacked high with new-mown hay.

Trucking proved a boon to loggers, who had previously been dependent on precarious and often jerry-built railways. Above is a five-ton chain-driven White truck in service for a logging company in Black Diamond,

Two fur-clad motorists pose in front of the White House on Pennsylvania Avenue in one of the numerous permutations of the Pope automobile. This one was a product of the Pope Motorcar Company in Toledo, Ohio, which was forced into bankruptcy in 1907. The Pope Toledo had been marketed as "The Quiet Mile-a-Minute Car."

the tone of the first presidential address delivered after the war, Charles Manly's in 1919. Most presidential addresses up to and including Charles Kettering's the year before had been pretty much alike—self-congratulation on behalf of the Society for one thing or another, often an admonition to those who seemed to be dragging their feet on standardization. Manly's was quite a departure: Though he himself was in aeronautics, he addressed himself largely to automobile engineers and their failure to fulfill their obligations to the public—not devoting enough attention, for example, to such growing problems as air pollution. Within a few years, expressions of concern about automobile safety were also becoming commonplace within the Society.

Professional Training

Manly was typical of SAE leaders who came from aeronautics, who usually had engineering degrees. The first and most famous of these was Elmer Sperry, a graduate of Cornell University. As the industry was still in its infancy, however, many key players did not: Glenn Martin, active in the Society for a dozen years, and Orville Wright both acquired their technical knowledge in the field.

Although the number of college-trained engineers among the first generation of automakers was surprisingly large—especially considering that any sort of engineering degree was still somewhat rare—men who headed SAE's Automobile Division had more often than not learned their skills working with machines in shops of one kind or another. As for those executives who were formally trained, most came from MIT; three presidents in a row, 1922–1924, were MIT grads, as were three of the six top SAE officers in 1924. After 1925, Rensselaer Polytechnic Institute and Worcester Polytechnic Institute were also represented, and more and more leaders and rank-and-file members had postgraduate degrees. By 1927 all but one of the top officers were university trained, including two from the University of Michigan, which by then was beginning to challenge MIT for the number of men reaching the upper echelons. Later, the University of Wisconsin became the one most heavily represented in SAE leadership.

Heading for a family outing at the seashore in a Model T, circa 1915.

Following spread: In a 1920s publicity photo, three women pose with a Packard Harveycar against a classic western backdrop. At the time, Packard's reputation as a luxury car was second to none, though as the market shook out at the end of the decade Packard was ranked only as a "major independent," along with Hudson, Studebaker, Nash, and Willys-Overland.

Broader Vistas

On the occasion of the Society's twenty-fifth anniversary in 1930, historical articles appeared, not only under SAE's aegis but also in *American Machinist* and *Automotive Industries.* These made it clear that the Society was already shifting from being just a standards organization toward being just as much a catalyst, a nerve center, an information exchange, and—in a phrase that later became commonplace—a vehicle for lifelong learning. Still, there was much change yet to come: As an article published on the Society's seventy-fifth anniversary put it, Fay, Coffin, Souther, and their pioneering SAE cohorts "would scarcely recognize" the SAE of 1980, perhaps foremost because of the level of cooperation it had fostered internationally as it became an organization with a worldwide reach.

1403

2

Sharing Information and Ideas

Why on earth do you need to study what's changing the country? . . .
I can tell you what's happening in just four letters: A-U-T-O!

ROBERT S. LYND AND HELEN M. LYND, *Middletown* (1929)

Opposite: Packard worked to maintain its luxury image in the latter 1920s with eight-cylinder models such as this Phaeton, and, beginning in 1928, eights for its smaller cars as well. Texaco, above, sold SAE Standard motor oil in cans that could be carried along for on-the-road replenishment.

A classic Margaret Bourke-White photo from the 1930s depicts a Kansas wheat farmer at the wheel of his Farmall and a man silhouetted against the sky while operating a combine.

As the first pressurized airliner, Boeing's Stratoliner of the late 1930s was capable of avoiding rough air by climbing above twenty thousand feet.

THE OBJECT OF THE SAE is to advance the Arts, Sciences, Standards, and Engineering Practices connected with the design, construction and utilization of self-propelled mechanisms, prime movers, components thereof, and related equipment, to preserve and improve the quality of life.

SAE CONSTITUTION AND BYLAWS

Andrew L. Riker, SAE's distinguished first president, felt confident that the word *automobile* was sufficient to cover "any self-propelled vehicle running on the land, in or under the water, or in the air." After a few years, however, certain members—especially those engaged with the rapidly evolving technologies of aeronautics—started to agree with Elmer Sperry that there was a word more properly descriptive of the Society's concerns: *automotive.* SAE became the Society of Automotive Engineers, Inc., in February 1917.

The new name was emblematic of a major reorientation as the Society moved into its second decade. In 1916 both the Society of Tractor Engineers and the American Society of Aeronautical Engineers merged with SAE; these reflected the primary interests of some members almost from the beginning. Likewise, the National Gas Engine Association and the National Association of Engine and Boat Manufacturers merged their standards work with SAE's, and soon the Society had five second vice presidents: for motor cars, tractors, aviation, marine engineering, and stationary internal-combustion engineering.

This separation of responsibilities lasted until 1929, when it was superseded by an arrangement of vice presidencies for each of eight activities: Aircraft, Aircraft-Engine, Diesel-Engine, Motor-Truck and Motorcoach, Passenger-Car, Passenger-Car Body, Production Engineering, and

Transportation and Maintenance Engineering. To these were later added Engineering Materials, Fuels and Lubricants, Tractor and Industrial Power, and in 1945 Air Transport. That same year, Aircraft-Engine Activity became Aircraft Powerplant Activity to keep in step with what would become known as the turbojet revolution.

Like the Wright brothers, Glenn Curtiss started in the bicycle business. He later moved into motorcycles with lightweight engines, and then he put such engines in airplanes controlled by ailerons rather than the wing-warping used by the Wrights—the first being the June Bug *in which he took wing on July 4, 1908. By 1914 Curtiss was the leading aircraft manufacturer in the United States.*

Taking Wing: SAE Becomes Aeronautical

The imagery of flight figured in Roger Bacon's thirteenth-century *Dreams of Art and Nature* and was foretold graphically by Leonardo da Vinci. Though dreams of powered heavier-than-air flight became commonplace during the nineteenth century in Europe and the United States, there is no doubt about who first fulfilled such dreams: two bicycle mechanics from Dayton, Ohio, Wilbur and Orville Wright. (This is quite unlike the invention of the automobile, for which one can argue on behalf of quite a number of inventors.) The Wright brothers were shrewd about practical matters but also well versed in technical literature and sophisticated about theory. This combination of attributes was the basis for the successful flight of the Flyer at Kitty Hawk, North Carolina, on December 17, 1903.

It is worth recalling that 1903 was the same year that Edward Tracy Birdsall began his campaign to establish the organization that became SAE. Interesting, too, that one of the Society's future leaders would be Charles Manly, who was given credit for the first powered flight by the Smithsonian Institution. Indeed, the Smithsonian, which was headed by Samuel P. Langley at the time, asserted untenably that Manly had made the first powered flight in Langley's Aerodrome a few weeks before the Wrights—this to the considerable displeasure of Orville, who years later as the surviving brother sent the Flyer to the Science Museum in London for display rather than the Smithsonian in Washington.

Genuine understanding of what the Wright brothers had accomplished at Kitty Hawk was slow to penetrate popular consciousness, and the skepticism was not put to rest until they staged a well-publicized demonstration for the U.S. War Department at Fort Myer, Virginia, in 1908. Not long afterward, there were two other news events along the same lines. Teddy Roosevelt went aloft in a Wright biplane, and the Wrights' major competitor, Glenn Curtiss, won the ten-thousand-dollar prize offered by a New York newspaper publisher for flying from Albany to New York City nonstop. Ensconced in their Manhattan offices, SAE brass surely noticed.

Soon, the phenomenon known as "air-mindedness" started to resemble a messianic religion, and pioneer manufacturer Jack Northrop declared that the airplane signaled a "glorious future of understanding and brotherhood." This was belied by a dawning realization of the airplane's utility in warfare—in 1917 Congress appropriated the unheard-of sum of $640 million to supply the allies with twenty thousand warplanes—but its potential benefits to humankind seemed nonetheless bounteous. Mail was being transported by air, for example, even before SAE embraced aeronautical pioneers Elmer Sperry, Glenn Curtiss, Orville Wright, and Glenn L. Martin. At the Society's annual meeting in 1921, long before any significant passenger traffic had developed, Martin delivered a paper on "Aerial Transportation as a Business Proposition."

The $640 million Congress set aside for warplanes helped U.S. manufacturers make some fifteen thousand planes—about a third of them by Curtiss—but only fourteen hundred were shipped to Europe. Here, in 1917, pilots of the 148th Aero Squadron clamor into action.

By then the international aeronautical engineering specifications program was well established. In October 1916 SAE had developed the first specifications—for the length of the metric

The workforce in manufacturing plants was warned about possible disastrous consequences if their attention to their duties was not up to snuff.

sparkplug and for a "tapered shaft-end for mounting propellor hubs"—in cooperation with Britain and several other nations. Henry Souther, now in charge of standards for aircraft engines, also reported on a definition of terms for engine rotation: "*normal* and *antinormal*, in place of the terms which have formerly been used and have caused confusion because they have not been used uniformly." By 1961 SAE aerospace standards would number one thousand, and the total in 2005 is more than sixty-two hundred.

All Realms of Human Mobility

The June 1916 meeting of the SAE Council is regarded as the most important in the entire history of the organization—it is said that this was when SAE "came of age." It was with this meeting that the Society pledged to extend its ideal of the free exchange of ideas and information into all realms of human automobility. This included off-road equipment—farm machinery at first, then machinery for earthmoving—and watercraft, but especially aeronautics. An unpublished history of the Society describes the 1916 meeting as thus:

> *The meeting room was electric with the presence of other groups of engineers who sought guidance from SAE. Famous inventor Elmer A. Sperry, Bion J. Arnold, and H. E. Coffin—representatives of the Industrial Preparedness Committee of the Naval Consulting Board—stated that they had come "to confer with the Council with the end view of having aeronautical engineers join and participate in SAE." The National Advisory Committee for Aeronautics was also on the scene, represented by H. C. Dickenson of the National Bureau of Standards and Henry Souther of the Aviation Section, Signal Corps, United States Army. They were interested in setting up better ways of dealing with aircraft technology.*

The chairman of the Naval Consulting Board, none other than Thomas Edison, had suggested the possibility of establishing an entirely new organization of aeronautical engineers. But Elmer Sperry and his cohort believed that they had a better idea, and it was immediately accepted by SAE leaders, many of whom agreed with Sperry's assessment of aeronautical technology's astounding potential. SAE's ideal of information exchange had expanded beyond the auto industry into a particularly significant field of engineering endeavor, one whose elaboration of design and construction principles was evolving at breakneck speed.

Elmer Sperry

The son of a lumber merchant from Cortland, New York, Elmer Sperry graduated from Cornell University and then founded the Sperry Electric Light, Motor, and Car Brake Company in Chicago, specializing in arc lamps and dynamos. Later he formed several more companies, experimented with battery-powered automobiles and diesel engines, and became the key figure in integrating aeronautical engineering into the activities of SAE. His vast range of inventions (he held more than four hundred patents) included the electromechanical gyroscope for stabilizing ships and aircraft, the gyrocompass, and the gyro airplane stabilizer. In 1928, two years before his death, he sold his Sperry Gyroscope Company to North American Aviation, today part of technology giant Unisys.

U.S. AERIAL MAIL SERVICE

The Liberty Truck

The War Department called on SAE for drawings for this truck, also known as the Heavy Duty Class B, in August 1917, and a team of engineers began its work on September 1. Fifty men worked twenty-four hours a day on the design, and construction of two prototypes was completed the next month. For road testing, these two started out from Rochester, New York, and Lima, Ohio, respectively, on October 10. They rolled into Washington, D.C., and received a personal greeting from President Wilson on the fourteenth. SAE standardization: a crucial contribution to the war effort.

Previous spread: In 1925, the inaugural day for U.S. Aerial Mail Service between Cleveland and Chicago, mechanics ready the biplane as a postal service vehicle arrives with pouches of mail. By turning the carriage of mail over to private contractors, the Air Mail Act, which was signed into law on February 2, marked the birth of commercial aviation in the United States.

The Liberty Engine

The new symbiosis bore fruit at once. In response to the secretary of the navy's request that the Society coordinate development of a standardized engine for military aircraft, in 1917 SAE's Elbert Hall and Jesse G. Vincent drew up preliminary designs for a four-hundred-horsepower liquid-cooled engine. Called the Liberty, the engine design was then submitted to machine-tool firms and aircraft-engine manufacturers—and to automobile manufacturers as well, because the Liberty had to be "a producer's engine," and nobody knew state-of-the-art production techniques as well as U.S. automakers.

By Armistice Day, Ford, Packard, Duesenberg, and Pierce-Arrow were among the automobile manufacturers that had turned out a total of 24,498 Liberty Engines for General Pershing's Allied Expeditionary Force and for the U.S. Navy, as well as for America's allies. SAE thereby made it plain to everyone involved in the war effort that this Society "held the key to unlocking the engineering potential of the aircraft industry."

At that fateful 1916 meeting, when aircraft and other engineering specialties were merged with automobile engineering to form automotive engineering, the Society's primary ideal was affirmed to lasting effect. As SAE secretary W. C. Redfield remarked, "There had been altogether too much separation in the past. Now, I hope the day of getting together has begun, and that the process of getting together will go further and further." In the nearly ninety years since, getting together has indeed gone far.

New Professional Activities

By 1922 SAE had a president from Autocar, Benjamin B. Bachman, a key figure in the transition from solid-rubber truck tires to pneumatic, and in 1925 it had a president, Harry L. Horning, who had been a moving spirit in the Society of Tractor Engineers and the National Gas Engine Association. But specialists with discrete forms of engineering expertise did not necessarily walk right into SAE's councils of power. Although SAE's constitution was amended in 1910 to invite

The day before completing his round-the-world flight on September 28, 1924, Lt. Howell H. Smith checks the engine of the Douglas World Cruiser Number 2, named Chicago. *Remarkable feats like this one helped the aerospace industry shake off its reputation as being unsafe.*

those involved with vehicles on "land, water, or air," and an aeronautics branch had been advocated in *S.A.E. Transactions* since 1911, many members continued to believe that airplanes were just too dangerous. Likewise, diesel powerplants, first noted in SAE annals in 1909, had a negative prognosis for their practicality in vehicles because they were just too heavy. Gasoline shortages during World War I sparked some interest in compression-ignition engines, but it was not until more than a decade after the end of the war that Diesel-Engine Activity found its own place on the Society's organizational structure.

Sometimes there was discord about nomenclature. Should the name be *freight wagon, industrial vehicle,* or *motor truck*? Should it be *motorcoach* or *motorbus* or just *bus*? What about self-propelled railcars with internal-combustion engines, such as the McKeen, particularly prevalent on Union Pacific and Southern Pacific branch lines? Even with the cornerstone of SAE's activities the question

Boss Kettering

The most famous of SAE's presidents—and perhaps the most illustrious auto engineer in history—was Charles F. Kettering. Born in Ohio in 1876 and a 1904 graduate of Ohio State, he worked for National Cash Register (NCR) until 1909 when, in partnership with Edward A. Deeds, he established Dayton Engineering Laboratories Company, or Delco. The device that made him famous was the electric starter—an invention whose technology lay in his experience at NCR—but his other inventions are legion. Two years after selling Delco to General Motors in 1916, he served as SAE's president, and two years after that he became president and general manager of GM Research, whence came shock absorbers, leaded fuels, automatic transmissions, safety glass, and Freon. Kettering retired in 1947, the same year that the so-called Kettering engine was announced in an SAE paper. The overhead-valve wedge-head V8 debuted in the 1949 Cadillac, and by the time "The Boss" died in 1958 such engines had become virtually the standard powerplant for American automobiles.

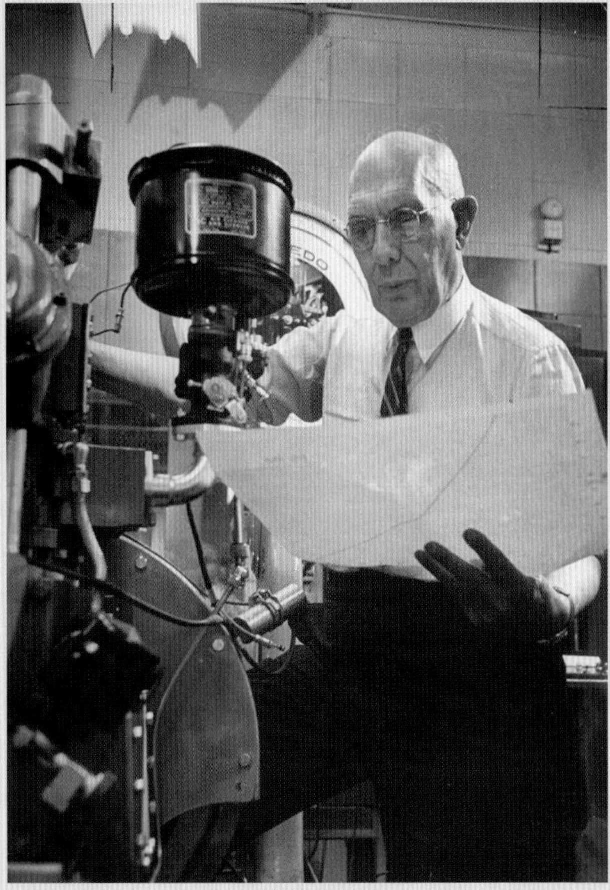

Above: "The Boss" pores over a dynamometer chart. At left, Kettering poses with his electric self-starter, the invention that made him an automotive legend.

A youngish Kettering is seen here in what was for him a rather unusual worksite, a chemistry lab.

Although the members of SAE were familiar with cars like this 1912 Model T, S.A.E Transactions *was surprisingly astute in predicting the future of the automobile industry.*

arose, would it be *motor cars*, *automobiles*, or *pleasure cars*? For a brief time, *road locomotives* even had its partisans. The word *pleasure* denoted something luxurious and thus more readily taxable, and so it was *passenger cars* that became accepted terminology.

As engineering specialties, it was passenger-car motors that first attracted the most attention, then transmissions, suspension, and brakes. The first SAE paper on materials was delivered at the Society's second meeting, in 1906, by the redoubtable Thomas Fay, who pleaded for "better materials than we are able to secure." But materials did not become a matter of general attention until later. Though no paper concerned with body design was presented until 1914, that paper was remarkably prescient:

A far cry from the old days: Autos had evolved just as envisioned in S.A.E Transactions. *With its hard top, nonprojecting lamps, and concealed spare tire, the Ford Mustang had succeeded the Ford Model T as the world's most popular Ford.*

> *The owners of the future will demand a car that will be comfortable to ride in at all seasons, that can be changed from open to closed with a minimum of time and trouble, and in which all accessories and spare parts will be covered and protected and easily kept clean. Bodies will look odd to us as we experience the evolution; first, the concealment of the extra tire, then perhaps the elimination of the usual folded top, then nonprojecting lamps, etc. Finally we will wonder why formerly we covered a car with so many trappings.*
>
> *S.A.E. Transactions*, 1914

Production engineering as a technical specialty was first mentioned in a 1910 paper that referred to those "directly connected with the daily output of motor cars and the economical operation of the factory." Measures aimed at reducing on-the-road expenses and delays—that is, increasing operating efficiency—began to attract explicit attention around 1920, as the number of autos

Henry Ford, right, and his son Edsel are seen here with Ford's first quadricycle, dating from 1896, and a 1924 Model T, the ten-millionth Ford to be produced.

worldwide approached ten million. SAE's president in 1921, David Beecroft, remarked, "It is just as essential that [autos] be kept at a standard of efficient operation and scientific application as it is that we go on building more."

One segment of the automotive engineering community was focused on fuels and lubricants as early as 1907, but the turning point was the establishment in 1921 of Cooperative Fuel Research, a joint activity of SAE, the American Petroleum Institute (API), the National Automobile Chamber of Commerce (NACC), and the National Bureau of Standards (NBS). Funding came from the API and NACC, the NBS was the investigative arm, and administration was the responsibility of SAE. Not until 1933 did Fuels and Lubricants officially join the ranks of the Society's professional activities. Tractor and Industrial Power Equipment followed two years later. Why were these and other significant specialties not included until relatively late in the Society's history? *The SAE Story*, published on SAE's fiftieth anniversary, suggests an answer:

> *The Society, having the best interests of the membership in mind, cannot move too fast, nor too slowly. It must guard against being overly influenced by the enthusiasms of the few, and be just as careful to recognize when an interest has expanded to assume an importance already recognized as specialized. . . . Even as the SAE grew in breadth by recognition and inclusion of allied interests, it has always singled out and emphasized the common denominators of automotive interest.*

Facilitating Progress

The process of cross-fertilization has always been central to the Society's activities. "Specialization," wrote a contributor to the fiftieth-anniversary analysis, is "a necessity in the modern world, but, at

Since the establishment of Fuels and Lubricants within SAE, fuel specialists have been able to work directly with members of related fields. One auxiliary industry was over-the-road tanker trucks, as shown above.

the same time, it tends to narrow the professional man's interest, and if allowed to run its course, it can defeat its purpose." Hence SAE has continuously enabled members to become conversant with fields other than their own: fuel specialists and engine specialists have enjoyed a rich synergy, for example, as have design engineers and materials engineers.

There was a similar influence in the way that standards were conceived—as "recommended practice," as a means of orderly development, but never as immutable. Never as a straitjacket, as some people in the industry initially feared, assuming that there would be Taylorist claims of "one best way to design an automobile." But standards as essential guideposts? Definitely so, first in the realm of nomenclature. Andrew Riker once lamented, "It would be a great plan to get at some definite names for the various parts of the automobile, because, as we all know, not every manufacturer and dealer calls the same parts by the same names." By 1914 a Research Division of the Standards Committee was in the works, the aim being not to tackle new problems but to arbitrate differences within subcommittees. What was the ideal size of tap drills to minimize

The Women of SAE

Treasurer of the Bantam Ball Bearing Company of Bantam, Connecticut, Nellie Scott was elected to an associate membership of SAE in 1920, the first female member. Amelia Earhart participated in several meetings, and the Society proclaimed in 1932 that it was "justly proud of those women who have the attributes and qualifications to become its members." Fifty years later, some engineering school classes were 30 percent women. By the late 1970s Women Engineers Roundtables were well established at SAE Congresses and SAE Off-Highway-Vehicle meetings, and by the end of the twentieth century, the Society had thirty-eight hundred female members.

Above: 2000 SAE president Rodica Baranescu, second from left, poses with three engineers from ARAL Research Center in 1999. At right, Amelia Earhart smiles from the cockpit of a Pitcairn autogyro—an aircraft with both a free-wheeling overhead rotor and a conventional engine at the nose; Earhart then worked for Harold Pitcairn as a publicist.

Earhart in front of her candy-apple Lockheed Vega, in which she soloed the Atlantic in 1932 in the record time of fourteen hours, fifty-six minutes. Here she poses not in aviator's leathers but in a skirt and heels.

Rosie the Riveter is an iconic figure usually associated with shipbuilding during World War II, but here three women work in aircraft fabrication in 1918.

breakage? How should printed forms for enumerating various elements of auto performance be standardized?

After World War I, a Research Committee (and later a Research Department) was established independent of the SAE Standards Committee. The Research Committee was headed by Hobart H. "Doc" Dickenson—a key participant in the June 1916 meeting—on leave from the National Bureau of Standards. Dickenson's policies would endure: there would be no encroachment on corporate research and development; rather, the committee would serve as a clearinghouse for information from industrial and university laboratories and would see to the information's dissemination in SAE's *Journal*, a monthly publication that was an integral benefit of membership.

Publications

Though the *Journal* publicized news of the Society and its members, it was first and foremost an engineering magazine that could be trusted to carry the latest technical information and reflect changes across the full spectrum of activities in which SAE members were involved. The direct successor, in 1917, of the *Bulletin*, which had been founded in 1911, it was the most important of SAE's publications. But it was not the first: *S.A.E. Transactions* predated even the *Bulletin*, having debuted in 1906. When the papers presented at annual meetings grew too voluminous, *Transactions* took on a new role. After World War II, the *Journal* became the place where members could find abridged versions of every paper presented at annual meetings; *Transactions* became a collection of the year's superior SAE technical papers. It was a delicate balancing act, of just the sort that the Society had already proven itself adept.

Take, for example, the question of what got printed in the *Handbook*. At its inception called the *Data Book*, the *Handbook* was published at the same time as the *Bulletin*, first in loose-leaf form (as was the *SAE Aeronautical Standards and Specifications*). The *Handbook*'s content—an annual compilation of the year's standards for ground vehicles and aerospace—was determined by the Technical Board, with the advice and consent of all the technical subcommittees. For the *Transactions* and *Journal*, it was (and is) the publications committee and its various reader committees—one for every activity field—that serve as anonymous referees.

The Manly Memorial Award

Charles Manly, SAE's president in 1919, first attained fame in 1903 as the engine designer and pilot for Smithsonian Secretary Samuel P. Langley's Aerodrome, in which two attempts were made to attain human flight just before the Wright brothers succeeded. While he was the Society's president, he worked to foster the industrial acceptance of standards and to further the work of the sections. Later, he became chief inspector for Curtiss Aviation, and when he died in 1927 an annual award was set up in his name. Inaugurated in 1928, this medal is awarded to the author of the best paper relating to theory or practice in research on—or design and construction of—aerospace engines or their parts, components, and accessories.

An illustration of Charles Manly's attempted flight in 1903. An SAE award was set up to celebrate Manly's pioneering work in aerospace.

Sturdy Foundation

The sharing of information and ideas through publications as well as meetings and sections has propelled SAE's associated industries forward year by year, though not always at a steady pace from one industry to another. Just as World War I precipitated a burst of energy, the two decades between the wars marked a transitional period in automotive history. Trucks moved more and more freight that had formerly gone via rail. A wave of airline mergers resulted in the creation of corporate names still familiar today—United, American, Northwest, Delta—and competition spurred the development of aircraft better suited to flying the mail. Soon these began to incorporate seats for passengers, as many as ten in a Trimotor, Ford's entry into this market. As new air-cooled powerplants from firms like Pratt and Whitney opened the door to commercial aircraft capable of cruising faster than a hundred miles an hour, more paying passengers showed up: 175,405 of them in 1929, 3.5 million in 1941.

Though automobiles aimed at the luxury market did not disappear from urban showrooms even in the depths of the Great Depression, low-priced autos came to be regarded as a necessity of

Opposite: Looking across Madison Square Park toward the spired New Life Insurance Company building in the mid-1920s, the streetscape is dominated by trucks, taxis, private autos, and a double-deck bus.

United Air Lines, Boeing's partner in United Aircraft, first flew the all-metal ten-passenger 247 in commercial service on February 8, 1933, and eventually the entire year's production went to United. One of the first 247s, unpainted to save weight, is seen here loading passengers at Boeing Field in Seattle.

everyday life in the United States. In addition, commercial motor carriers, mechanized farm and construction machinery, motorcycles, and even private planes contributed to a perception of a golden age in the progress of automotive technology. At every SAE meeting, as one firsthand account puts it, "the discussional hopper overflowed with technical treatises on air-cooled engines, mechanics of lubrication, metallurgy, new materials . . . mechanical vs. hydraulic brakes, disc wheels vs. wood . . . manifold design, fuel distribution, engine balancing."

Balloon tires first necessitated the redesign of steering systems, and then engendered the reengineering of entire motor vehicles. In 1906, SAE members had regarded 1,500 rpm as maximum operating speed for internal-combustion engines; by the 1930s they thought of 4,500 rpm as moderate and 7,000 rpm as appropriate for racing engines.

SAE's placement service was busy finding engineers for jobs and jobs for engineers. Members shared the assumption that special requirements demand special talents and that automotive engineering, being a dynamic enterprise, will take many engineers to jobs in many distant places.

Toward an Automotive-Engineering War

The mobility industries blossomed and grew after World War I in large part because of the near-constant stream of new ideas from engineers: diesel power, radial engines, flight test instruments, all-weather navigation, refrigerated trucks—the list goes on and on. Even throughout the Great Depression, SAE members kept presenting exciting papers on everything from aluminum alloys to independent suspension. When war in Europe loomed again, SAE's wartime experience a generation earlier proved invaluable as the Society hurried to lend its expertise to the Allied effort. The SAE War Engineering Board was established in 1939 to evaluate problems defined by the military and to assign committees of specialists to ascertain prompt solutions. Three years later, SAE's directors authorized formation of a War Activity Council to focus the efforts of the board. As an SAE history written shortly after the war puts it, "Plans and operations were geared to the speed of the wheel and the wing, calculated to the power of the engine, conceived for inhumanly complete destruction. Concepts of war were cast in global proportions. Automotive engineering became

SCHLOSS
P. FLORIN
ROBBINS
PROKESCH
MADISON
SQUARE
HOTEL

P&H

Left: Mass-producing twenty-eight-ton tanks in 1942 at the Chrysler factory in Detroit.

Previous spread: A leading-lady of the 1920s and 1930s, Lelia Hyams, poses with her imposing black Packard sedan, the last word in luxury.

part of military strategy, production of combat. Preparations for battle no longer involved the comparatively simple conditioning of men. It called for complex global readying of design, manufacture, delivery, and utilization of materials. Manpower was measured less by numbers than by capacity for making and using machines of war."

World War II was an automotive-engineering war, based on the techniques of mass production that had been pioneered and perfected in the automobile industry. For such a concerted effort, the fraternal spirit of SAE was ideal. The cooperative association of engineers involved in SAE and the competition among engineers in industry both enhanced preparedness. With lessons learned during World War I still fresh in mind, coordination was all that was required to gear the Society's automotive-engineering capabilities and the industry's productive capacity to the needs of mechanized conflict.

Women at work at Douglas Aircraft in Long Beach, California. World War II engendered demands on the aircraft industry that enabled it to adapt automobile production methods to suit its own needs.

Under the chairmanship of Chrysler's James C. Zeder, the War Engineering Board started out with thirty-six projects. Eventually it was engaged in more than four times that many, involving the U.S. Army, U.S. Navy, Quartermaster Corps, Ordnance Department, Signal Corps, War Production Board, Office of Defense Transportation, Bureau of Ships, and Army–Navy Aeronautical Board. It advised on issues ranging from enhanced gas and tire mileage to methods of conserving critical materials: aluminum, copper, chromium, iron, and, of course, rubber. It taught the army how to operate motor vehicles in the Arctic and taught the navy how to overcome difficulties caused by torsional vibration in marine diesels. For the Ordnance Department it compiled data on helical, spiral, leaf, and volute springs. It reverse-engineered captured enemy equipment. It developed protective packing and shipping materials for vehicles and components. Ultimately, the War Activity Council involved thirty-two hundred SAE members, who contributed more than fourteen hundred documents to the basic technical literature. More than 3.5 million copies of *SAE*

The Wright Brothers Medal

The most venerable of the SAE special awards, this medal was initiated by the Dayton Section and became the responsibility of the national society in 1927. It is presented to the author of the best paper on aerodynamics, structural theory, research, construction, or operation of airplanes or spacecraft presented under the Society's auspices.

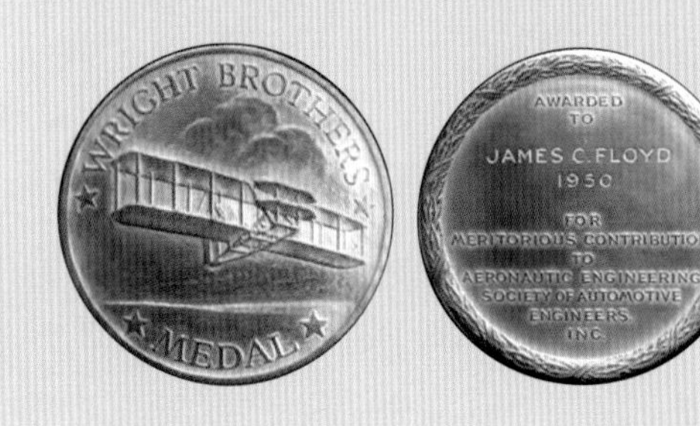

Aeronautical Standards and Specifications were published. The Society's membership grew from 5,855 in 1940 to more than 12,000 in 1945.

The complete tally of SAE contributions to the war effort fills dozens of pages. SAE's wartime service validated—fully and for all time—the sort of cooperative engineering efforts that SAE's founders had envisioned forty years before. It also led to the formulation of plans for transferring SAE's wartime operating strategies to peacetime activities. Even before VJ Day, the SAE board appointed a Postwar Advisory Committee, assigned "to study all phases of SAE operations, looking to a smooth transition from war to peace and the development of a dynamic postwar program . . . to consider how best the Society could profit from the experience it had been accumulating, and, incidentally, make provision for continuing work with the military."

Among the committee's recommendations, the most crucial was derived from the experience of the War Engineering Board: creation of a technical board to absorb the General Standards, General Research, and Engineering Relations Committees and administer all work of twenty-two technical committees. The Technical Board was soon acting as the central administrative agency for projects that came from industry and serving as a crucial liaison between military and industrial engineers.

Advancements in Aerospace

When the war ended, every branch of automotive engineering—automobiles, airplanes, heavy duty—appeared to have a boundless future. As the value of door-to-door service became more obvious and the design of tractor-trucks and semitrailers improved dramatically, long-distance trucking made tremendous inroads against the railroads. By the mid-1950s more than ten million trucks of all kinds were registered in the United States. By the early 1960s more intercity travelers were riding in motor buses than on railroad trains. At the same time, commercial turboprops and turbojets heralded a revolution in air transport that would all but vanquish the passenger train and consign the ocean liner to history.

This revolution in air travel had its roots in wartime experience. By the time of Pearl Harbor, aircraft production was already three times what it had been in 1929, and comprised a lion's share of a defense budget that was racing toward one hundred billion dollars. At the 1941 National Aircraft Production Meeting, which was devoted to issues of coordination, General Hugh S. Johnson—former chairman of the Moline Implement Co. and one-time head of Franklin

Roosevelt's National Recovery Administration—told the assembled executives and engineers that "Industry has got to do its stuff. That stuff can't be any ordinary business-as-usual, workaday performance such as we see in peace. *You have got to perform miracles!*"

As domestic and foreign orders multiplied for both military and commercial aircraft, avoiding differences in specifications between one plant and another became a critical issue. By 1942, SAE's Aircraft Materials Division had made certain that common standards prevailed all across the industry. SAE engineers were also instrumental in converting nonessential manufacturing operations to wartime needs and helping improve production methods, methods that remained in place after VJ Day.

The SR-71 Blackbird, one of the most innovative aircraft to come out of Lockheed's Advanced Technology Division, also known as the Skunk Works.

In the realm of aircraft propulsion, tremendous technical advances had been made during the war, with fighter planes designed around piston engines of unprecedented horsepower. Soon, however, both the aircraft industry and the newly created Air Force were pushing SAE to develop the technology of jet propulsion. In the vanguard was the XP-80, the new single-engine jet fighter, whose design team was led by SAE Fellow Clarence L. "Kelly" Johnson, working in Johnson's later-famous "Skunk Works" at the Lockheed plant in Burbank, California. By the late 1940s, with performance records being set and reset over and over, SAE members had been instrumental in assuring the future of jet power. In 1950, SAE's Aeronautics Committee established new divisions for developing military standards for materials, engines, and propellers, as well as equipment and special projects.

SAE involvement in this technological revolution went far beyond military airpower. While piston-engine airliners had been vastly improved since their introduction in the 1930s, by mid-century it seemed obvious that the future lay in a new form of propulsion, or, rather, new forms of propulsion. SAE engineers were as yet unable to agree on the relative advantages of turboprops

SAE Celebrities

While SAE has had members from all realms of engineering, it has had many members who were also celebrities: Orville Wright, Glenn Curtiss, Charles Lindbergh, Boss Kettering, Henry Ford, and Jimmy Doolittle, to name just a few. Will Rogers once was a keynote speaker at an annual meeting. More recently, SAE membership rolls have included Dan Gurney, the legendary racer and team owner; Andy Granatelli, of Indy racing fame; and Jay Leno, *Tonight Show* host and auto enthusiast.

Above: During the dedication of the Hall of Aviation at the Franklin Institute in Philadelphia, Orville Wright poses alongside Amelia Earhart, with just a wheel of the plane in which Earhart soloed the Atlantic visible overhead. The date is December 17, 1933, the thirtieth anniversary of Orville's first flight at Kitty Hawk. At left, Charles A. Lindbergh is seen with Spirit of St. Louis *on the occasion of its presentation to the Smithsonian Institution in 1928.*

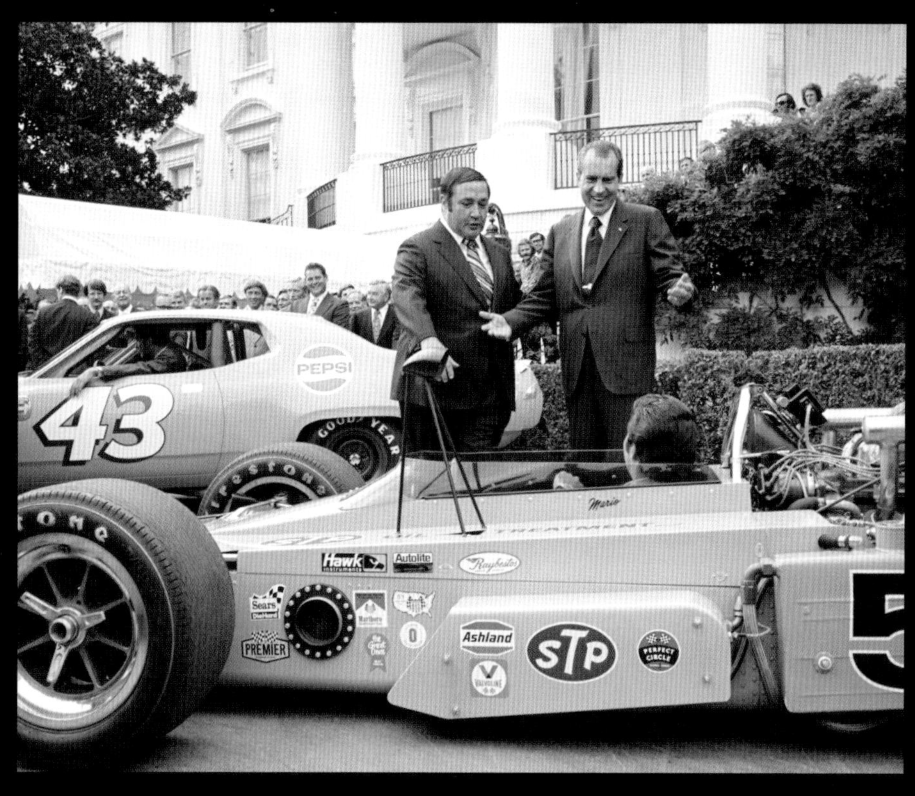

Above: During a White House reception honoring celebrities from the auto racing world in September 1971, Andy Granatelli stands by as President Richard Nixon grins at Mario Andretti, in the cockpit of Granatelli's STP special. Among many other things, Dan Gurney, right, was the best American Formula 1 driver ever; he is seen here in 1970 at about the time he retired from the cockpit. Jay Leno, far right, is probably the world's most famous car buff.

Opposite: On July 16, 1959, with Neil Armstrong, Buzz Aldrin, and Michael Collins aboard, Saturn V *lifts off NASA launch pad 39A at the Kennedy Space Flight Center. The three astronauts are headed for the moon.*

and turbojets, and at the 1950 national meeting agreed to devote equal time, energy, and brainpower to both. William Littlewood, an SAE member who was vice president of engineering at American Airlines, remarked that "we must be prepared for some false starts and mistakes . . . the sooner we get started, the sooner we will get results."

By the mid-1950s, SAE engineers were engrossed in seeking solutions to problems of engine vibration, excessive fuel consumption, and the effects of high altitude on crewmen and passengers. SAE made available to industry a vast range of new standards and recommended practices—an example of the latter being Recommended Practice 580, "Cockpit Visibility Requirements for Commercial Transport Aircraft," which was accepted for all new jet transports introduced after 1956. In 1961, the SAE Aerospace Council formed a committee to evaluate the problem of jet noise, and that same year also marked the publication of SAE's one-thousandth aerospace specification.

At San Diego's Lindbergh Field, a Boeing 707-123B starts its take-off roll down Runway 27. When this photo was taken in 1980, more than seven hundred 707s were in service worldwide.

A few years before, at a dinner celebrating SAE's fiftieth anniversary, Arthur S. Klein, professor of aeronautics at the California Institute of Technology, had made a prophetic remark about "the new generation of aircraft engineers." These engineers, said Klein, "will make our future vehicles safer, more efficient, and useful, and may—by means of some as yet unknown powerplants—enable us to escape into space." The world was right on the verge of the space race, with SAE's members beginning to ponder satellites, space stations, and spaceflight. Dr. Wernher von Braun told a meeting of SAE's Detroit Section about the multistage booster called Saturn, designed for launching a probe that could make a soft landing on the moon. In the realm of suborbital aeronautics, SAE engineers were beginning to turn their attention to supersonic transport (SST), and throughout the decade of the 1960s there was rarely an issue of the SAE *Journal* that did not mention the burgeoning SST project. In the end, however, Great Britain and France would finish designing and constructing the Concorde and then put it into operation, while—to the lasting sorrow of the many SAE engineers who had devoted so much to the project—the American SST would never take wing.

When introduced in 1954, the Caterpillar D9 was larger than any other tractor on the market. By 1959, when this photo was taken, the diesel-powered bulldozer has assumed its classic configuration.

Off-Highway

Back on terra firma, however, the area of technological enterprise then termed *Off-Highway* and now formally called *Heavy Duty* by SAE was flourishing as never before, as the farm tractors permuted into specialized applications with crawler treads, hydraulic accessories, and mechanical appendages. SAE was directly involved in the work of the Nebraska Tractor Test Laboratory, whose comprehensive reports were supplied to farm organizations, university agricultural departments, and governmental agencies worldwide. The test laboratory dated back to 1920, only four years after the SAE embraced the Society of Tractor Engineers. The impetus had come from concerns about the poor performance, durability, and dependability of farm tractors, of which there were at least 160 manufacturers in the early 1920s. The Nebraska legislation stipulated that "a stock tractor of each model sold in the state be tested and approved by a board of three engineers under State University management." As the system evolved, eventually the Tractor Test Board worked with two advisory committees, one comprised of farmers and tractor dealers, the other representing industry and serving as a subcommittee of the SAE Tractor Technical Committee.

At the same time that tractors were evolving in their agricultural applications, variations of tractor technology were also becoming a vital tool in mining and especially in construction, as the development of the interstate highway system—the largest civil works project in history—fostered an enormous demand for earthmoving machinery. Beginning in 1956 the design of the interstates, with four and six lanes, cloverleaf interchanges, and restricted gradients and curvatures, entailed earthmoving operations of unprecedented magnitude. As the system reached beyond forty thousand miles, the cost soared towards one hundred billion dollars and transformed the industry that became known generically as "yellow steel."

By the mid-1960s, machines such as these three Euclid R-20s and the 72-40 were indispensable to open-pit mining operations.

Hydraulic controls became the norm. The introduction of the integral wheel tractor-scraper marked the beginning of the end of the power shovel. Engines, tires, and everything else grew to gargantuan dimensions; practical limitations in size were reached by the late 1960s, at least for machines that needed to be moved from one site to another, but mining machinery continues to grow even today. As the interstate system neared completion in the early 1980s, what was once a torrent of contracts for the earthmoving-equipment industry and for road-building contractors dwindled away to a trickle, and there was a severe shakeout in both lines of endeavor.

Where work remained to be done, it was not in new construction but in repairing and upgrading existing highways, particularly increasing the number of lanes. This sort of work required, as William Haycraft puts it in his history of the earthmoving equipment industry, "more flexible methods and created demand for hydraulic excavators, articulated dumpers, elevating scrapers, and mid-sized equipment rather than the heavy crawlers and large scrapers used in new construction."

To scan current issues of *SAE Off-Highway Engineering* is to find articles not only on such flexible midsize machinery but also on equipment ranging from a forty-four-foot-high mining machine dubbed by its manufacturer, LeTourneau, as "the world's largest front-end loader" to a

Seen in 2004 at the Buckskin Mine near Gillette, in Wyoming's Powder River Basin, is a LeTourneau front-end loader with twelve-foot tires and a bucket capacity of forty-six tons. Though a comparatively small operation, Buckskin annually sends twenty million tons of low-sulfur coal to eastern power plants.

neat little 9000M farm tractor from Stery in Austria, with a four-cylinder turbocharged engine, two-speed transmission, and curved glass to reflect outside noise downward. One may also glimpse the future in the form of "autonomous off-highway vehicles" such as John Deere's pioneering apple-orchard sprayer, vehicles for which the development of SAE standards will prove a crucial step toward practicality. One notable aspect of the orchard sprayer is that that it lacks the ROPS (rollover protective structure)—no driver, no danger—that is now all but standard on off-highway equipment and whose design has been a cooperative effort between the Construction Industry Manufacturers Association and SAE, with its never-ending development of performance standards.

Americans on the Road

By the time President Eisenhower signed the Federal-Aid Highway Act authorizing the interstate system, SAE membership stood at almost twenty thousand, and the horizons for the automobile seemed boundless. Annual production in the United States hovered around 6.5 million, the Big Three automakers controlled two-thirds of the world market, and their hegemony seemed assured. But automobiles had mixed portents. In 1954, New York City's traffic commissioner declared that, "most drivers are not sufficiently trained nor physically able to handle the horsepower placed at their disposal." For the first time the United States became a net oil importer. Though they were bigger, heavier, and more powerful, rarely if ever had automobiles been as poorly designed as they were in the late 1950s, and Ralph Nader's book *Unsafe at Any Speed* was only the best-publicized example of a deluge of muckraking attacks on the industry.

Nothing symbolized the interstate highway system, with its enormous appetite for space, as much as the cloverleaf interchange. Here, in Los Angeles—the city most closely associated with urban superhighways—Interstate 10, also known as the Santa Monica Freeway, soars across Interstate 110, the Harbor Freeway.

As auto ownership reached one of every three people in the United States by the 1960s, the sociologist Vance Packard proclaimed that "the primary function of the motorcar in America is to carry its owner into a higher social stratum." There were still, however, significant innovations reaching the market: Buick offered the first V-6 engine in 1960, Pontiac the first transistorized ignition system in 1962, Oldsmobile the first front-wheel drive in 1966, and Ford began making radial tires optional in 1968. GM built its hundred-millionth vehicle, Ford its seventy-millionth. But it was not the best of times for the U.S. automobile industry, which was rapidly losing ground to European and Japanese imports, nor was it the best of times for SAE, whose membership numbers remained flat for much of the 1960s and 1970s. With the expansion of global markets, it became clear that SAE would need to expand globally as well.

Beginning in the latter 1970s growth came quickly. SAE's focus on education and international initiatives, along with its new understanding of engineering's pivotal role in society, would lead to its transformation into a vital progressive force worldwide.

SOUTH
SOUTH
EAST
TRUCK
BY-PASS
TRUCK
13
13
308

NORTH NORTH

WEST TRUCK BY-PASS

308 13 13

3

Laying the Groundwork for the Future

SAE derives its personality, character, direction and continuity from the dynamic involvement of its members in the engineering world around them.

Joseph Gilbert, 1967

Through its rigorous and thorough standards, SAE's three core constituencies, including heavy duty, left, and aerospace, above, have laid the groundwork for a safe and efficient future.

Ford's new plant in Flat Rock, Michigan, known as Auto Alliance International, turns out six new models of 2005 Mustangs on two vehicle platforms. Ford's robotic devices put into effect SAE's challenge to use manufacturing methods that offer better value to the customer.

An upsized 2004 Toyota Prius, the make and model that most Americans associate with hybrid gasoline engine/electric motor power plants. The hybrid car allows engineers to improve fuel efficiency as well as to protect natural resources.

IN 1980 THE SOCIETY OF AUTOMOTIVE ENGINEERS celebrated its diamond jubilee. The Society could reflect on an enormous number of accomplishments with great pride. But there were also major challenges in the engineering world—technological, environmental, and societal—and in a special seventy-fifth-anniversary publication the Society identified nine:

- *To balance mobility and environmental needs*
- *To make more efficient use of available energy and develop new energy sources*
- *To enhance the degree of safety associated with mobility*
- *To use existing materials more effectively and to develop satisfactory replacements for materials in limited supply*
- *To employ new designs, materials, and manufacturing methods that result in products yielding better value to the customer*
- *To develop advanced design concepts and techniques that result in products meeting user needs and societal needs*
- *To achieve a better understanding and appreciation of the engineering approach to meeting society's needs*
- *To bring about an increased interaction with other disciplines (such as political science, law, and life sciences) as a means of developing socially acceptable products*
- *To educate engineers to cope successfully with change and sociotechnological problems*

A 2004 Toyota RAV4 at the moment of impact during a crash test conducted by the Insurance Institute for Highway Safety.

From the perspective of the Society's centennial twenty-five years later, all these challenges are worth another look, but some seem particularly significant: those concerning education, communication, outreach, and interaction with other disciplines. Consider the essay by James Eagan, the marketing manager for new materials at MPD Technology Corporation, who predicted in 1980 that polymers, and also fiber- and graphite-reinforced composites, would see increasing use in conveyances of all kinds. To confirm Eagan's perspicacity, one has only to glance at the recent SAE Special Publication from the 2003 World Congress, *Advances in Plastic Components*, and see that Eagan returned to a theme—cross-fertilization—that has been fundamental throughout SAE's history:

> *Good communication between related engineering disciplines is a must. The materials engineer is the cornerstone in meeting the materials challenge, so he must make certain that he communicates with his fellow engineers, and vice versa. Also, the materials engineer should to some degree become a design and manufacturing engineer, and these engineers should in turn become to some degree materials engineers.*
> *The SAE Challenge: Freedom through Mobility*, 1980

Or consider the exchange of ideas and information in a more general sense. Robert Brown, executive vice president for corporate development at Eaton Corporation, began his SAE seventy-fifth-anniversary essay by noting that automotive engineering had reached a crossroads. Having met the challenges of "spearheading freedom through mobility," it now needed to address a wide range of issues such as pollution control and enhanced safety, areas where the automotive engineering community would be compelled to interact with not just technical experts but also "individuals, groups, and organizations exerting a broad spectrum of influences on complex technical issues." This interaction he foresaw taking many forms, but for SAE it was partly a challenge to educate the general public, to impart an understanding of not only what automotive engineers

At the Honda plant in Suzuka, Japan, workers attend to the final assembly of 1981 models. Honda manufactured its first automobiles in 1962 and within little more than twenty years had become the leading foreign producer in the American market.

had achieved historically but also the finite bounds of engineering's capacity to solve social problems. Thus, Brown concluded,

> *as interaction becomes the password of the twenty-first century, automotive engineering . . . must relate to the forces of change, anticipate its impact and get involved in shaping change. It will mean changing our priorities and rearranging our allocations of time and effort to meet the challenge squarely.*
> *The SAE Challenge*

"Automotive Engineering Knows No National Boundaries"

One essential change SAE had begun to make was to realize the importance of international cooperation. SAE president Harry Chesebrough back in 1960 remarked, "Automotive engineering knows no national boundaries," a fairly revolutionary idea at the time. An SAE delegation had visited Europe before World War I, and in the 1970s those international engagements became frequent. How could they not, with activities in far-flung factories and boardrooms influencing affairs in the United States almost daily? In 1980 Japan surpassed the United States in auto sales.

A Material World

"The use of any material begins with understanding how it behaves," writes historian Robert Friedel, who continues:

> *The world offers up an enormous variety of substances, and human modification yields even more. Choosing among these materials may be simple or straightforward, or it may be the result of a complex series of tests and decisions. The final choice will reflect values and circumstances that may be economic, political, cultural, and aesthetic. But often the most basic element in the choice of material is physical—will the substance behave as it must in order to be useful, safe, practical, attractive, or stable?*

The predominant materials in autos, as in all conveyances—the "stuff from which they are made"—have inevitably changed over time. In 1907 Alanson P. Brush, who had been an engineer for Buick and Cadillac, struck out on his own with the Brush Runabout. Even before Henry Ford, Brush sought to court a mass market by means of a rock-bottom price; the last Runabouts, called Liberties, sold for $350. But more interesting than Brush's anticipation of Ford's marketing strategy was the means by which he sought to keep his manufacturing costs down. Although wood was not uncommon in early-day autos—wooden spokes were all but universal—Brush used oak, hickory, and maple for not only the wheels of Runabouts but also the frame and axles, and other components as well. A Liberty weighed only nine hundred pounds and was not very durable. Yet, while Brush was behind the times with his use of wood in highly stressed components, he was also ahead of his times in many ways: his crankcases, for example, were aluminum.

Below: The Brush Runabout, produced between 1907 and 1912, incorporated several innovative technologies, such as a planetary gear-set with an internal-tooth annular gear. These gears were later used in overdrive units and in the automatic transmissions that were commonplace by the time Mr. Brush died in 1952. Despite the extensive use of wood, a Runabout was mostly steel, just as steel predominates in the manufacture of today's autos, despite the increasing use of artificial polymers.

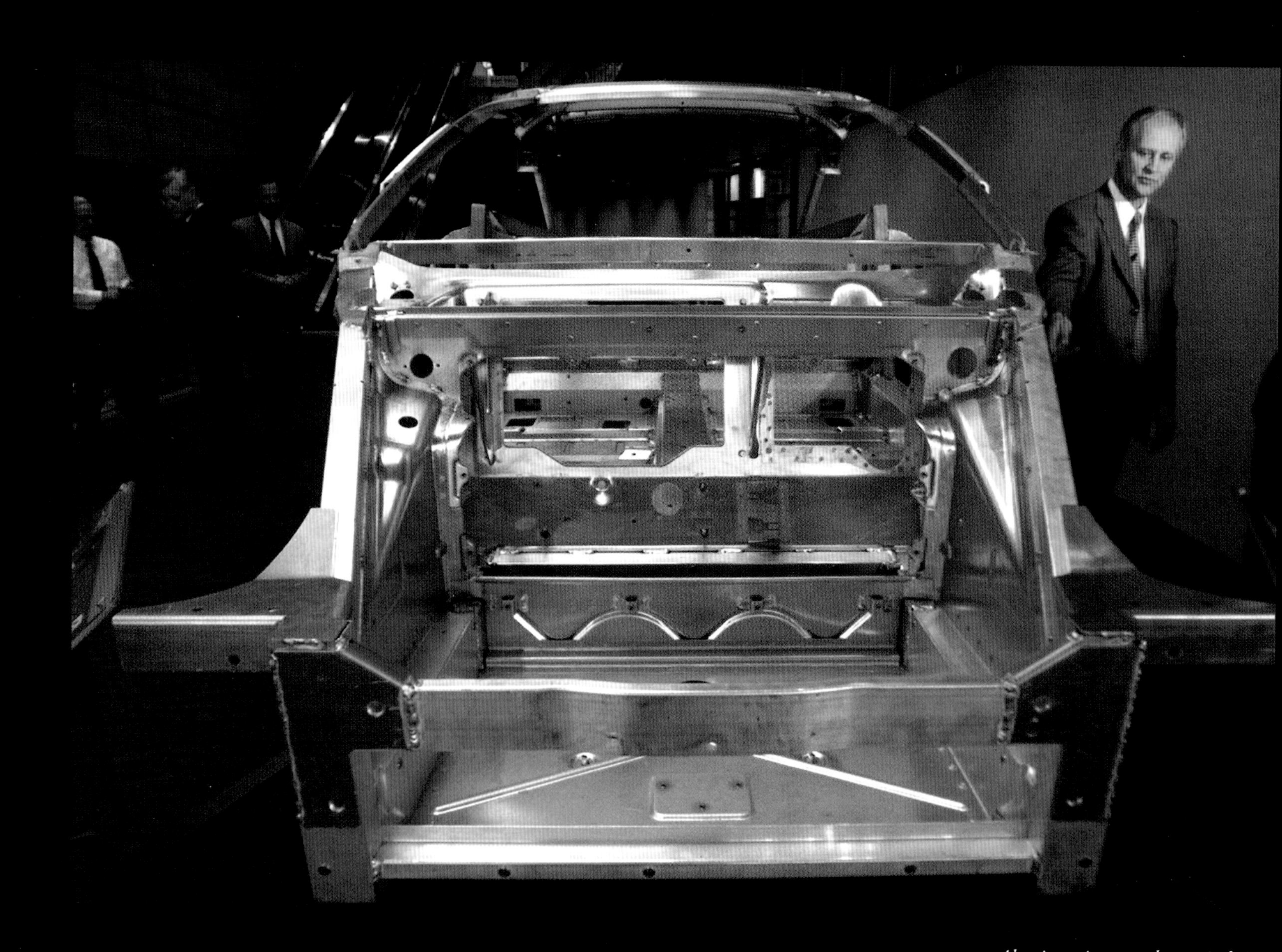

Aluminum is not used extensively in most autos, though there are exceptions. Here, at Alcoa's Pittsburgh headquarters, Miloslav Novak points to the aluminum chassis of a Ferrari 360 Modena, which also features an aluminum block and aluminum suspension components—and a price tag far into six figures.

DODGE MAGNUM
HEMI
5.7L V8
ENGINE
FOR GASOLINE ENGINES

Opposite: Americans increasingly prefer large and powerful vehicles, and since 2000 the number of light cars and trucks with high-displacement V-8s has grown annually. Here, a man admires the 5.7-litre Hemi Magnum in a Dodge Truck.

In December 2002 General Motors, partnered with Shanghai Automotive Industry Corp., purchased the Yangtai Bodyworks in Shandong Province, thereby becoming the only foreign automaker with four factories in China. Here, newly employed Chinese workers look on intently during a technical training session.

Three years later, of the eleven firms that produced more than a million vehicles, only three were headquartered in the United States, while four were Japanese, two were French, and one each were West German and Italian. Of some thirty million cars manufactured worldwide in 1983, more than three-quarters were produced outside the United States. (Compare that with 1923, when all but about 400,000 of the 4 million cars produced worldwide were made in the United States, or even 1953, when it was 7.3 million made in the United States out of a world total of 10.5 million.)

During the 1980s, design variations for different domestic markets looked like they were becoming a thing of the past. A world car such as the Escort might be assembled in the United States or the United Kingdom, in Germany or Portugal, with components obtained in a dozen other countries. In the 1990s the international trend to downsize cars was significantly reversed, as vehicles aimed at the American market grew to outsize proportions and the horsepower race that had been put on hold for most of a generation resumed once again.

Still, an auto (or truck, tractor, or airplane) of the twenty-first century is likely to have parts from Europe or Latin America or South Africa, no matter its place of assembly. Though they may have a different terminology, specifications for components in Boeing and Airbus products are often identical. Worldwide standards in the auto industry are as crucial as ever, because the market is still growing. The People's Republic of China has a market for tens of millions of autos, where not long ago there were next to none; in 1980 there was one passenger car for every 18,673 persons. Since 1990 new highway construction in China has averaged thirty-seven thousand kilometers per year, and a nationwide grid second in extent only to that of the United States is predicted by 2020. This naturally puts stars in the eyes of automakers; Volkswagen has already invested five billion dollars in China.

World Headquarters

From its very beginning, from the days of Edward Birdsall and Andrew Riker, SAE was headquartered in New York City. The offices were in several locations over the years; by the 1970s the offices were at Two Penn Plaza, atop Penn Station and Madison Square Garden. This was very expensive real estate, New York was a very distracting place, and the Society's management had come to the conclusion that, after seventy years, a move was in order.

Detroit? SAE had long had a presence in the Motor City, but for its headquarters? Too car oriented for SAE's aerospace members. Washington, D.C.? Too much danger of diminishing SAE's hard-won reputation for independence. Well then why not Chicago, Atlanta, Dallas, Los Angeles, San Francisco, or Denver? The choice was Pittsburgh, and the Society's general manager, Joseph Gilbert, explained the rationale of the Ad Hoc Relocation Committee to the staff and membership in these words:

> *The Committee reasoned that the basic objective is to locate SAE's principal offices where the work of the Society can be carried out most efficiently and effectively. Therefore, high on the list of location considerations is overall economics of headquarters operations. This includes the cost of owning or renting or operating the offices, staff salaries, and the cost of staff travel. It was also deemed important that headquarters have ready access to the kinds of business services it needs for its daily activities . . . . SAE headquarters, in the view of the Society's officers, is in reality a workshop where the Society's business functions are performed.*

So, how did Pittsburgh stack up among the twenty-one cities considered? It is midway between Washington and Detroit. The airport was nearby. Pittsburgh is the location of many of the corporations central to the industries with which SAE is involved: United States Steel, PPG, Alcoa, Westinghouse, Koppers, Bayer, National Steel, Gulf Oil. And Rockwell International, involved in all of the industrial segments central to SAE's affairs, played a major role in attracting the Society to western Pennsylvania, or more precisely, Warrendale. Colonel Rockwell was an early-day SAE member and his company made significant contributions to the SAE building fund that was under the charge of Arch Colwell, a former SAE president.

Planning for the move began early in 1972 and ground was broken on Thorn Hill in Warrendale on November 7, 1973. The move was completed in the fall of 1974. Today, Warrendale is known as SAE World Headquarters, while the Detroit offices—much larger than in former times—are known as SAE Automotive Headquarters. The Society's chief operating officer Ray Morris remarks that this setup is "very close to ideal."

An artist's rendering of the newly rebuilt visitor's entrance to SAE's

Technicians at Energomash in Moscow discuss RD-180 engines slated for shipment to the United States for installation as the first stage of Lockheed Martin's Atlas 3 rocket, exemplifying the international cooperation that SAE has encouraged for decades.

And all this has had a marked impact on SAE. In the mid-1970s, the Society's membership had not grown for ten years, holding steady at thirty thousand members. Amid oil shocks and concerns about gridlock and environmental degradation, books such as *The Death of the Auto* were appearing. Fortunately, the dire situation wasn't permanent for SAE: by the mid-1980s the Society's membership was edging close to fifty thousand, and in the late 1990s, with a new SAE Global Development Office in the works, it topped eighty thousand. There were SAE members in almost eighty countries.

The first real spark of interest in international relationships dates to the presidency of Leonard Raymond in 1959. Raymond toured much of Western Europe, hammering at a single theme wherever he went: communication barriers—between countries, between companies, even within companies—desperately needed bridging. By the 1970s SAE was engaged in varied and increasingly constructive modes of interchange with engineering communities in other lands, and engineers outside North America considered publication of a paper by SAE as a mark of highest prestige. In just one year, between 1978 and 1979, the number of Society groups involved in projects with an international dimension jumped nearly 100 percent. The reasons were clear: inflation, decentralization and denationalization of manufacturing, changing export–import relationships, and increasing support for uniform modes of regulation. Today, as the worldwide market becomes more interconnected, SAE's worldwide influence is growing as well.

Password of the Twenty-first Century

Education is the heart of every technical challenge, but in addition to learning about technology, engineers must also learn "to cope successfully with change and sociotechnological problems." This

On a June day in 2003, drivers on interstate 75/82 in Atlanta are warned of a Code Red Smog Alert, the result of unsafe ground-level ozone concentrations. SAE has been working steadily to help improve fuel efficiency and perfect alternative fuels in order to rectify sociotechnological problems such as air pollution.

was how Phillip S. Myers, a distinguished professor at the University of Wisconsin and past president of SAE, put it in his contribution to the seventy-fifth anniversary forecast. Myers wrote that:

> *engineering education is really a liberal education for both men and women. It requires a solid foundation in both humanities and social studies, as well as a thorough knowledge of chemistry, physics, mathematics and engineering principles. It demands understanding awareness of environmental problems. . . . Engineers must understand the economics, both national and international, of the problems they are solving . . . [and work toward] development of new energy sources, pollution reduction, conservation and utilization of natural resources, space travel, information processing and urban renewal.*
> *The SAE Challenge*

In other words, engineers need a broad understanding of sociology. Suburbanization in conjunction with inner-city decay? Enhanced individual mobility along with the demise of the extended family? Ecological awareness casting a spotlight on technology's environmental downside? The engineer opens the door to prodigious change, but responses are always constructed

clean
natural gas
CNG
MAR
California
2005

At a demonstration for the California Air Resources Board in 2004, Seth Seaborg of Environmental Vehicle Outfitters displays the "EVO Limo," a Suburban converted to compressed natural gas in the interest of reducing ozone depletion and global warming.

through political processes. To scholars concerned with the relationship of technology and cultural context, Myers's insight was as fundamental as the free exchange of ideas and information was to automotive engineers.

Another major area of potential growth was in the realm of lifelong learning. There were many opportunities for this already, none so rich as those provided by SAE through its meetings and publications. By the 1970s SAE had shifted from a primary focus on standards to an equally powerful focus on its role as catalyst and nerve center. In 1980 SAE was also facing the challenge of unprecedented change in its political and economic context and of changing needs necessitating innovative educational techniques. It was clear that education needed to be a top priority.

Members of the New Jersey Institute of Technology's 2003 Mini Baja Team pose proudly with their work in progress. The Mini Baja is emblematic of SAE's emphasis on imparting pride and pleasure to educational activities.

Generations

While the Society must never fail to meet the intellectual needs and serve the aims of seasoned engineers, it must also provide the same benefits for newcomers to the profession. The evidence suggests that its educational strategy has paid off almost since the Society's beginning. One of the first students elected to junior membership was Chester Ricker in 1911, whose distinguished engineering career would last for more than fifty years and include a long stint as chief timer at the Indianapolis Motor Speedway. Ed Cole, another early junior member, became chief executive officer of GM.

Student branches date from 1915, when there were eighty-seven student members. Today well over four hundred student branches provide a major arena of contact with real-world hands-on engineering; the awards program for faculty, established in 1964 by former SAE president Ralph R. Teetor, aims at improving the design and effectiveness of engineering curricula. In the latter years of the twentieth century, however, nothing has put students, practicing engineers, and engineering professors in such an unusual and fruitful interactive relationship as the building, evaluating, and testing of small special-purpose vehicles for the Society's Mini Baja and Formula SAE competitions. The competitions now extend around the globe as part of the SAE Collegiate Design Series.

To interest children in math and science, SAE has created the AWIM Challenges. Here, seventh graders work on Challenge 2, Motor Vehicles.

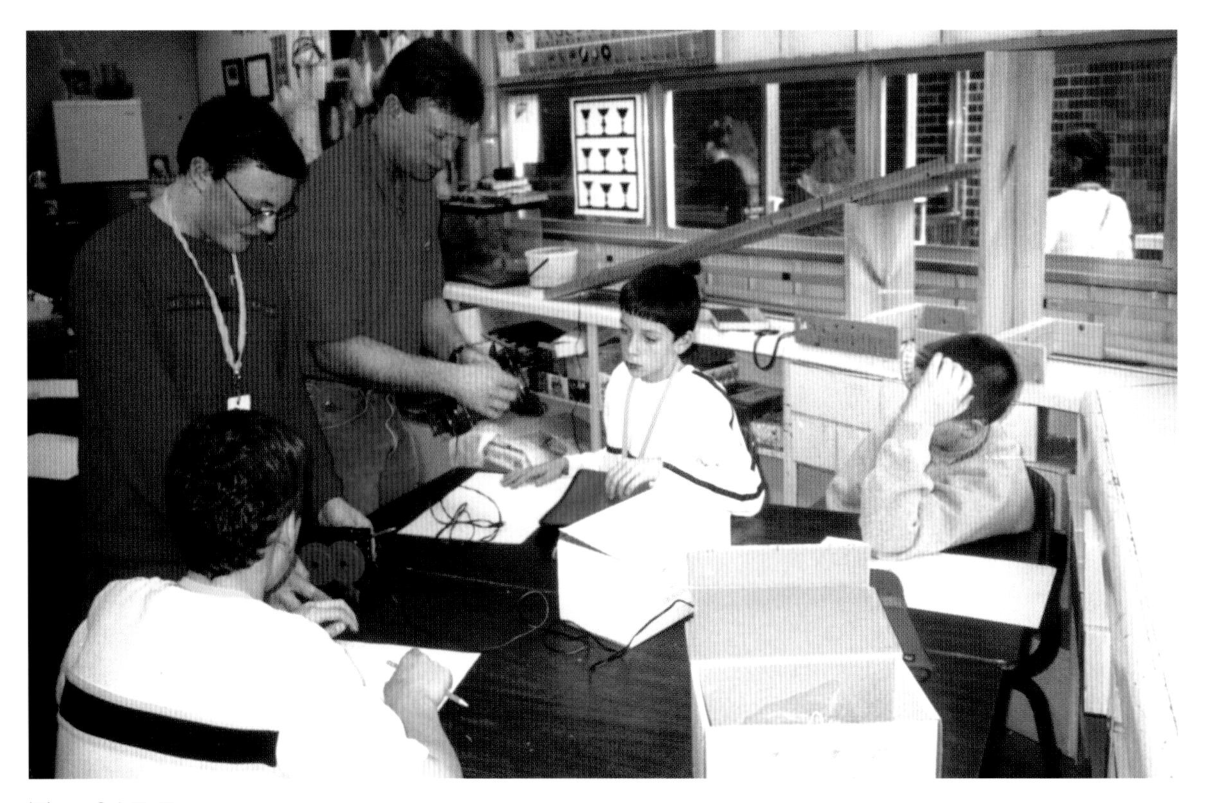

The SAE Foundation

In the mid-1980s, the SAE board of directors recognized the need to establish a mechanism for funding and promoting the advancement of education in math and science outside the normal framework of SAE's operations. The result was the creation of the SAE Foundation in 1986.

The initiatives supported by the SAE Foundation include:

- the award-winning A World in Motion program (AWIM) for grades four through ten
- numerous awards for both young and seasoned professionals
- undergraduate and graduate scholarships
- dynamic competitions offered through the Collegiate Design Series

The foundation promotes an interest in math and science in grade school and middle school students through the AWIM program by collaborating with professional engineers and scientists in schools. These mobility professionals work with teachers to mentor students in the various design contests that both educate and entertain students. Perhaps the most important effort of the

Three Magazines and *UPdate*

Today the Society's three magazines, *Automotive Engineering International* (successor to the *Bulletin* and now in its 113th volume), the newer *Aerospace Engineering* (established in 1981), and *SAE Off-Highway Engineering* (established in 1991), reach beyond SAE membership with cutting-edge technical articles and news of developments in their respective industries. In the early 1980s, Ray Morris, who became SAE's chief operating officer in 2001, began publication of *SAE UPdate*. This monthly publication features news about the Society's members as well as information of particular interest to members, such as schedules for the Professional Development Seminars, which are fundamental to the Society's program in continuing education. Since 1998 SAE technical papers and standards have also been available online.

SAE's publishing program has been wildly successful. About a quarter of the membership buys SAE publications, but the majority of sales are to nonmembers, men and women involved in automotive industries for whom SAE provides an invaluable product.

SAE's magazines below, keep Society members and nonmembers alike informed of the latest advancements in the automotive industries.

Michigan governor John M. Engler, addressing reporters at the 1996 SAE convention in Cobo Center, speaks of plans for an on-line Auto University and auto-industry job-retraining programs.

foundation is to build bridges between corporations and classrooms by bringing engineers, teachers, and students together so that they may all learn from each other.

The foundation funds several awards each year for both young and experienced professionals to encourage the advancement of various engineering professions and SAE programs that promote engineering.

Scholarships are critical to promoting the advancement of engineering, and the SAE Foundation funds several collegiate scholarships. Some date back to 1994 and others are recently established. One, the Women Engineers Committee Scholarship, established to encourage young women to pursue engineering degrees, was started in 2003.

SAE's many Collegiate Design Competitions are also funded through the foundation. These highly touted programs reach over five hundred teams with more than forty-three hundred competitors every year. They are open to undergraduate and graduate students, and in some cases, to high school students. They offer aspiring engineers the opportunity to transform theory into practice.

The foundation is supported by donations from corporations, SAE members and staff, private donors, and other foundations.

VISION 2000

As the SAE Foundation evolved during its first few years, it became clear that its fundamental responsibility was for educational activities. VISION 2000 was the name given to the Society's campaign to support engineering education at all levels, the aim being "to ensure that there are enough engineers to meet the needs of industry, education, and government in the next century"—this in direct response to the Carnegie Commission's eye-opening report titled "Nation at Risk." Programs that became part of VISION 2000 included the student design competitions; undergraduate engineering scholarships; the Industrial Lectureship Series; the program of graduate-student loans; the SAE Faculty Advisor Awards; the SAE Teetor Educational Award Program, aimed at strengthening the bond between engineering education and industry; and most important, the Engineering Education Board's program A World in Motion.

A World in Motion

SAE's pioneering elementary and middle school effort, A World in Motion (AWIM), was launched in 1990 with the specific aim of bringing a new style of pedagogy to physical sciences in grades four, five, and six. AWIM sent out thousands of classroom kits containing a video, teacher guide, and set of dynamic hands-on classroom experiments. As a catalyst for action at the community level, SAE members—often distinguished senior members—were tapped to assist teachers offering the program in designated school districts and to serve as mentors and role models.

AWIM was an instant success. By the end of 1991, thousands of teacher kits had been distributed throughout the nation, all the Canadian provinces, and parts of Mexico. Ten thousand more were on order. By 1996 the total number distributed was more than thirty thousand, and AWIM had received official recognition from the National Research Council. It had also become the curriculum enhancement of choice for the Johns Hopkins University New American Schools initiative, and was forming the basis for unique partnerships—in Milwaukee, for example, where corporations such as Delco and Rockwell had joined with local firms to bring AWIM to the Discovery World Museum and involve nearly fifty schools from the metropolitan area.

By 1997 AWIM had brought together more than fifteen thousand volunteers from SAE sections with nearly a million students at thousands of schools all over North America. For the most imaginative and effective use of AWIM materials in a classroom setting, SAE established the Lloyd Reuss Award—named for the former GM president who had chaired the SAE VISION 2000 executive committee. Thanks to active promotion, AWIM has become the nation's most comprehensive private initiative for promoting engineering education in elementary schools.

Janice Henderson

Recipient of the 1999 Lloyd Reuss Award, Janice Henderson is a fifth-grade teacher at Amerman Elementary School in Northville, Michigan. Henderson had used AWIM in her classroom for more than five years, during which time six volunteer SAE engineers had participated. Said Henderson:

> *The students became so engaged that they began taking their boats home and involving their parents. Parents then interacted with their sons, daughters, and friends. Strong partnerships began developing between home and school. I have never seen such enthusiasm for science exhibited in school before. It was great! I still have the parents and students stop me at local stores or soccer games and tell me what a great time they had.*

A World in Motion II: The Design Experience

Within a few years after the inception of the original AWIM program (soon renamed AWIM Challenge 1), SAE started to build on its success, this time with a $1.8 million challenge grant from the National Science Foundation. Designed by leading specialists in multidisciplinary curriculum development, AWIM Challenges 2 and 3 are eight-week exercises related to land, sea,

The SAE Doctoral Scholars Program

The aim of this program is to encourage top-flight engineering students to pursue the doctoral degree and then return to teaching at the college or university level. Funded by the Chrysler Corporation, it provides student loans each year for a three-year period, one year of which is subsequently forgiven for each year of teaching the doctoral scholar completes. Said, Richard Nelson, an MIT student and one of five new doctoral scholars named in 1996 (and bringing the total over the years to sixty-four):

> *Teaching is the key to tomorrow's future; outstanding teaching is insurance for its success. By allowing me to continue engineering studies at a prestigious university, the SAE Doctoral Scholars Program has enabled me not only to become a good researcher, but also to continue my dream of serving as an effective teacher.*

and air mobility technology for grades six through eight. It quickly won the endorsement of the National Middle School Association and the National Association of Secondary School Principals.

The scenario presented to students is this: a company called Mobility Toys, Inc., has issued a request for proposals and prototypes for a motorized toy vehicle. To respond, students first conduct market surveys and then design, build, and test a prototype using materials provided in a laboratory kit. AWIM Challenge 2 is a land-vehicle project for middle school students. During its first year, AWIM Challenge 2 reached more than eight thousand students throughout the United States and Canada; by 1998 this number had grown to eighteen thousand and by 1999 to nearly forty thousand. Challenge 3, where middle-school students design a glider and write an instruction manual to teach others on how to build the same glider, reached over ten thousand students in 1998–99, its first year.

Students learn for the first time what it is like to engineer a new product prototype. Both the land-vehicle and glider Challenges (and now a Challenge 4 involving electrical technology) entail working through a sequence of steps in a comprehensive new-product development process: setting goals; building knowledge; designing, building, testing, and finalizing a product; and making a final presentation. AWIM Challenges make engineering come alive for children, and help them see how dynamic it can be as a career. These Challenges keep with the lifelong learning ideals that SAE holds so dear, and are the surest way to keep future generations interested in engineering.

Extended Education

For much of history, engineering was learned in a similar way as the AWIM Challenges: hands-on at the worksite or in the shop. Not until shortly before SAE was born was there any systematic effort in the United States to link engineering to rigorous standards of higher education. How much has changed since then is indicated by SAE's fundamental orientation toward formal education, be it the Challenges or in university postgraduate training. Even after an engineer's formal training is complete, educational opportunities abound: an amazing range and diversity is to be

During an October 2001 test session at the Dunn County Recreation Park in Menomonie, Arlen Kruger of the University of Wisconsin-Stout SAE team takes flight in his Mini Baja car.

found in SAE's *Professional Development Catalog*, which outlines the content of hundreds of seminars on nearly a hundred topics for professional engineers.

Other evidence for SAE's commitment to lifelong learning can be found in its technical publishing catalog. Along with its growing list of general-interest publications, SAE's wide-ranging special-publication program features monographs on the cutting edge of technology. Publishing has been integral to SAE since 1906 and its influential body of work includes authored books, books comprised of SAE papers, and electronic publications in several forms. These publications help further the lifelong learning that is so essential to SAE.

Collegiate Design Competitions

The Society's first collegiate design competition, the Recreational-Ecological Vehicle (REV) contest, was launched more than thirty years ago, in 1973. This soon evolved into something different under the guidance of Professor John Stevens of the University of South Carolina. In the now-numerous REV contests, student-designed vehicles emulate those that compete in the Baja 1000 off-road race. Engineering and design teams attempt to create a prototype of an all-terrain vehicle that fits in the bed of a pickup, appeals to weekend off-roaders, and can be produced for

GKN
Fox Shocks
Patriot Steel
EAGLE

SAE

The Recreational-Ecological Vehicle (REV) Contest

Predating the Mini Baja competition by three years was the REV contest conceived by Dr. William Shapton in 1973, to have students design and fabricate a two-man amphibious ATV "capable of negotiating twenty-five miles of virgin Michigan forest and several hundred yards of deep water." The ATVs in the REV competition were one-offs costing as much as five thousand dollars, and some were masterpieces of automotive engineering. Even though this was not the sort of project that would stimulate the design of a mass-produced vehicle such as those designed for the Mini Baja, the REV series opened the eyes of the SAE to the value and virtue of such technical competitions.

Previous spread: The 2003 Formula SAE team from the University of Akron looks over its car. Formula SAE has become internationally renowned for bringing together the brightest engineering students from around the world.

less than three thousand dollars in a production run of four thousand. Briggs & Stratton—an intrepid supporter of SAE activities for many years—donates spec engines (eight-horsepower at first, increased to ten-horsepower in 2000), but the rest is up to undergraduates collaborating with working engineers, professors, and graduate students. Vehicles are judged on the basis of design (safety and cost) and performance on the basis of acceleration, handling, power, and endurance. Many student teams have ended up designing vehicles for the real thing, the Baja 1000.

Ten university teams participated in the inaugural Mini Baja at the University of South Carolina and Fort Jackson in 1976. Within two years, the competition had been separated into three regions, and in 1981 twenty-seven schools participated in the Southeast Mini Baja alone. By 1984 the number of teams competing in this one region had grown to forty-five. A Mini Baja competition was first held in Canada in 1985 and one is now held there every three years. Teams currently participate in Mini Bajas from all over the world, and as we move into the middle of the first decade of the twenty-first century, their numbers are increasing.

Formula SAE

The three regional Mini Bajas have also been joined by several other Collegiate Design Series events, including Supermileage (inaugurated shortly after the Mini Baja), Aero Design East and Aero Design West, the Walking Robot Challenge, and perhaps most notably, what is called Formula SAE, begun in 1981. Formula SAE is a competition for student members to design, create, and compete with small formula-style race cars.

In 1982, in hopes of luring entrants with modified Mini Baja cars, another category of competition was created for cars with Briggs & Stratton engines and suspension on all four wheels. The suspension rule remained in effect, though the spec engine experiment was later dropped—which opened the door to motorcycle powerplants, fuel injection, forced induction, and methanol. Teams showed up with rotary engines, aerodynamic wings, and even powered ground effects. To prevent design parameters from being stretched too far—as racers are wont to do—the rules makers eventually stipulated a maximum cost for Formula SAE entries.

SAE Supermileage Competition

Stressing innovative design with ultimate safety, this competition aims to produce vehicles of maximum fuel efficiency, and the results have been astounding. In 1997, Privateer, the creation of a team from the Technical University of Nova Scotia, set a mark of 992.59 miles per gallon, outdistancing the runner-up by 330 miles per gallon. Since then, the record has shot up to more than sixteen hundred miles per gallon. As with the Mini Baja, engines for Supermileage are donated by Briggs & Stratton.

The winner of the 2003–04 Supermilage Competition was this entry from California State University, Los Angeles, defeating thirty-nine other entries and vastly exceeding all previous records by attaining 1,615 miles per gallon.

Renee T. Sears

In 1989, Renee Sears, a graduate student at Cornell University, joined the Cornell Formula SAE team. Ten years later she still recalled her Formula SAE experience—spending every waking hour with her teammates—as the most worthwhile aspect of her university studies:

Through Formula SAE, I learned to balance priorities and to communicate clearly what I know and what I need, so that others can understand in ways that will be meaningful to them. I use those skills every day. . . . Participation in a team-based engineering project like Formula really tests your comprehension of the basics, and it gives you the opportunity to fail while you're still in a very supportive learning environment. Through Formula, I gained a sense of confidence in my own technical work that I did not get from working problems in textbooks.

Above: The champion 1998 Formula SAE entry from Cornell University drifts into a right-hand corner and, left, perpetuating Cornell's winning tradition is the champion 2004 entry.

In 1986 Formula SAE began to move and grow. Ford, GM, and Chrysler took turns as host of the race, after which a Formula SAE consortium was created with representatives from each of the Detroit automakers and also from the Society's Educational Relations Department. In 1995 eighty-four teams competed at the Silverdome in Pontiac, Michigan; in 1997 a British entry from the University of Leeds did extremely well; and in 1998 a spin-off competition, Formula Student, was inaugurated in the United Kingdom.

On September 13, 2004, the men above signed the documents that created the SAE UK affiliate. The SAE UK affiliate incorporates the Institute of Vehicle Engineers as well.

The SAE Foundation Cup—established in 1998 to recognize Formula SAE winners for excellence in engineering, design, teamwork, creativity, and project management—was first awarded to Cornell University, which has dominated Formula SAE along with the University of Texas. But at the same time, the field was becoming ever more diverse: The 2001 competition featured entrants from Japan, Mexico, South Korea, and Puerto Rico, as well as from the United Kingdom; the 2002 competition attracted 118 teams from eight countries on five continents.

"The future of automotive technology is here at Formula SAE," remarked the Society's 2004 president, Duane Tiede. And it is certainly true that Formula SAE provides opportunities for hands-on teamwork that is all but absent from the educational experience in other engineering disciplines. Some automotive firms almost exclusively hire students who have been on Formula SAE and Mini Baja teams, knowing that they already understand the hands-on aspect of engineering that would take years to learn as employees.

But Tiede had much more in mind. The future implies globalization. The history of Formula SAE and Mini Baja epitomizes the most significant trend of the past twenty-five years: rapid movement into worldwide realms. In accordance with this trend the Society again changed its name in 1980, becoming SAE International.

One World

SAE International initiatives are often focused on India, China, and the countries of the former Soviet Union, which together have about two-thirds of the world's population. They also have large contingents of engineers eager for state-of-the-art knowledge and growing consumer markets. During the 1990s SAE began signing memorandums of understanding (MOUs) with

In 1996, during a week-long Beijing exhibition of electric and alternative-fuel vehicles from eleven countries, a visitor eyes a car designed on Hainan Island that can be supplied with either electric, natural gas, or diesel power. SAE hopes to establish stronger ties with China in the near future.

organizations around the globe, "for two parties to achieve common objectives by some kind of informal agreement on a broad framework of cooperative actions." In addition, international sections were formed in Russia, Belarus, Egypt, Romania, Italy, Malaysia, Ukraine, Israel, and a dozen other countries. By 2000 more than twelve thousand SAE members—practicing engineers, professors of engineering, and students—resided in countries other than the United States and Canada. Efforts at expanding the SAE reach throughout the world were led by then-SAE executive vice president Max E. Rumbaugh, Jr.

In 1992 SAE took the ultimate step in international cooperation and formed its first affiliate—SAE Brasil, with engineers and executives from twenty Brazilian and multinational firms constituting a senior advisory directorate, and operations complementing the activities of other Brazilian automotive-engineering societies. SAE presidents Phil Mazziotti (1981) and Ed Mabley (1989) had begun laying the groundwork for the Brazil affiliate years before this. Members enjoy dual membership with SAE International and receive all membership benefits. After the SAE board defined globalization as one of two critical strategic issues, SAE International helped SAE Brasil host its First International Congress and Exposition in October 1992. Two years later Brazil staged a Mini Baja in São Paulo, and afterward two of the Brazilian teams traveled to the Midwest Mini Baja in Ohio. In 2000 SAE Brasil launched its own magazine *Engenharia automotiva e aerospacial.*

SAE's leadership believed that SAE Brasil would not be the only SAE International affiliate. It took a while for the second to materialize, but when it did it was in India, a nation with one of the world's two largest populations, whose automotive economy seemed to be bursting at the seams. SAE leaders had visited India as far back as 1969 and committed the Society to supporting Indian automotive engineers and students. Despite this effort, by 1995 SAE's Indian section had only two hundred members. That year SAE president John Leinonen initiated processes subsequently supported by succeeding presidents to travel to India with the purpose of establishing an affiliate there. The inducements are substantial: Affiliates have their own leaders and manage their own affairs but also are strongly assisted in establishing institutional and individual links with SAE International. On October 19, 2002, SAE leaders signed an MOU with SAE India leaders, and SAE had its second affiliate. By 2002 SAE India had fifteen hundred regular members and forty-five hundred student members distributed in sections covering all four centers of Indian automotive production.

Brian Whitesell

In the mid-1980s, Brian Whitesell served as crew chief and driver for Virginia Tech's Mini Baja team. The experience served both as a means of putting theory into practice and learning important lessons about teamwork. Said Brian, "Since the program was very young at Virginia Tech, it was still in the developmental stages and they were having trouble getting things accomplished. I learned a lot about leadership and making sure that things got done. It also cemented for me just how much I wanted to get involved in motorsports."

And did he ever. In 1992, Brian joined Hendrick Motorsports as the tractor-trailer driver for Jeff Gordon's NASCAR team. Two years later he was named team engineer, three years after that he earned Western Auto's Mechanic of the Year award, and in 1999 he was promoted to team manager, responsible for everything from construction to budgets and scheduling.

Right: Jeff Gordon pits during the 1998 Country Music Television 300 at the New Hampshire International Speedway, an event he won with Brian Whitesell as team engineer. Above: During practice for the 1999 NAPA 500 Winston Cup race at Martinsville, Jeff and newly-promoted Brian discuss Brian's reading of a sparkplug.

SURE DEPLOYMENT MENU

Tom Yager shows off the SmarTruck at the 2001 SAE World Congress in Detroit. A U.S. Army prototype, the fully armored SmarTruck is safely controlled inside the truck via touch-screen displays.

Indian engineers have since organized mobility conferences, workshops, and student competitions and have established an SAE India Foundation with aims and initiatives patterned on the SAE Foundation in Warrendale and the SAE Foundation Canada.

And now there is a third. SAE had long had a strong section in the United Kingdom's Midlands and a sister society called the Institute of Vehicle Engineers (IVE). The SAE section and the IVE merged in September of 2004, forming the SAE UK.

Both the IVHS America and SAE have been working for several years to construct an intelligent transportation system. Above is a 1995 prediction of intelligent auto controls, the "Cockpit 2000."

It seems unlikely that there will ever be SAE affiliates in nations that already have powerful and venerable engineering societies, such as the Verein Deutscher Ingenieure (VDI) in Germany. Where well-established national societies already exist, SAE's strategy has been to attract membership by offering access to its incomparable store of engineering knowledge; the SAE database is richer than those of all the other thirty societies in FISITA—the international federation of automotive engineering societies—put together.

In the People's Republic of China, whose potential automobile market is the largest in the world, SAE implements its Continuing Professional Development program by means of a joint agreement with Delphi at Tsinghua University in Beijing. Spend any time with SAE International's top leaders and you are bound to get wind of a global development plan with a focus on China as the keystone of an effectively linked "worldwide network of technically informed mobility practitioners." If student design competitions hold the future of automotive technology, then surely MOUs and affiliates hold the future of SAE International.

Intelligent Transportation Systems

As far back as 1991, SAE had published nearly a dozen monographs on intelligent vehicle–highway systems (IVHS; the original name for Intelligent Transportation Systems, or ITS) and become a founding member of IVHS America, a nonprofit corporation whose purpose is "to advance a national program for safer, more economical, energy efficient and environmentally sound highways in the United States."

The Society also established a program to coordinate ITS activities involving standards. Under the auspices of SAE's Cooperative Research Program, a coalition of universities, government

Boeing 737s in production at Seattle. By 1993, Boeing had sold more than three thousand 737s, more than any other airplane in the history of commercial aviation. Boeing is one of the many companies that benefits from PRI's internationally recognized audits.

organizations, and manufacturers began investigating issues such as ITS evaluation and liability. In 1994 SAE played a key role in the World Congress of Intelligent Transportation Systems, held in Paris and attended by three thousand delegates. The challenges of ITS to SAE are enormous, because they entail an inherent dilemma that the sociologist Rudi Volti describes thus:

> *The principal barriers to ITS deployment are likely to involve issues of cost and equity more than technical feasibility. Some critics, noting the sheer growth in the number of single-occupancy vehicles and vehicle trips, argue that a fully deployed ITS infrastructure will offer only temporary relief from congestion, and that these benefits will accrue mostly to drivers who can afford expensive in-vehicle technologies. . . . There are also concerns about whether ITS can be optimized to produce simultaneous benefits for mobility, air quality, energy conservation, and social equity. ITS is like an amplifier. It amplifies both what is good and what is bad about our transportation system, and extends the consequences to other domains, such as energy, environment, and land use.*
>
> *The Facts on File Encyclopedia of Science, Technology and Society* (1999)

The Performance Review Institute

The Performance Review Institute (PRI) was created in 1990 to advance the interests of the mobility industries (air, land, sea, and space) through development of performance standards and administration of quality assurance, accreditation, and certification programs.

An affiliate of SAE, PRI is a nonprofit organization with offices in the United States, the United Kingdom, and China. PRI's current staff of sixty performs three thousand audits annually. Rather than have various companies and the Department of Defense perform redundant audits, PRI does the only audit—a service to industries that saves millions of dollars. According to Ray Morris, the "aerospace industry looks at SAE as the standards developer and they look at PRI as the conformity assessment, but they look at the two together, with PRI as a value-added service to the industry."

The key products—better described as services—PRI offers to industry are Nadcap and the PRI Registrar. Nadcap is an industry-managed, consensus approach to OEM oversight of special process and product suppliers. Nearly all of the world's aerospace primes (OEMs) use Nadcap certification as a major component in their supplier oversight process. The PRI Registrar certifies organizations to a variety of management systems including ISO 9000, AS9100, and ISO 14000.

Left: In October 2004, Mario Dominguez powers his Ford-Cosworth through the chicanes at Surfers Paradise on the coast of Queensland, Australia. Below: Indianapolis Motor Speedway's Tony George, stopwatch in hand, follows a qualifying run for the 500.

The Motorsports Engineering Conference

SAE has been involved in its own forms of motorsports competition since the inception of the Mini Baja in the 1970s. But the Society began to move more directly into motorsports in 1990, when it included a motorsports session at its World Congress—"Design and Development of the Modern Indy Car"—with a panel of experts in engine and chassis design, tire development, fuels, safety, and specialty components. State-of-the-art engines and chassis were on display, as well as historic machines that had run in the Indianapolis 500 in years past.

So successful was this 1990 session that it led to the first SAE Motorsports Engineering Conference and Exposition, in December 1994. Organized by distinguished members of the racing community, the technical sessions addressed performance simulation and modeling, aerodynamics, fuels and lubricants, suspension, and structural analysis. Within a few years, the Motorsports Engineering Conference had evolved into one of SAE's major biennial events. By 2004 SAE had joined with Panoz Racing in sponsoring a seminar on suspension setups, and it was billing the December Motorsports Engineering Conference in Dearborn as "the industry's premiere high-performance technology event." An ever-growing international dimension was underscored by the keynoter, Max Mosely, president of the Federation Internationale de l'Automobile (FIA), the world governing body of motorsports. Panels were headed not only by Mosely, but also by NASCAR president Mike Helton, Indianapolis Motor Speedway CEO Tony George, and National Hot Rod Association president Tom Compton. A Young Engineers Panel was convened in full confidence that "the dream of every engineering student or young engineer is to land a spot on a team designing the next generation racer of high-performance car."

The conferences also had sessions that addressed historical epochs that went back to the very beginnings of SAE, such as Henry Ford's response to the Selden Patent suit, and there was a tour of the Ford Rouge Plant. Even though there is no aspect of automotive engineering that is more intensely focused on the future than the design of racing cars, there seems to be a satisfying symmetry in the high-performance world's most significant annual conference also devoting attention to a hundred years of history.

Above right: The signatures of everyone who worked on any given Panoz Esperante are engraved on a sheet of metal that is screwed to a carbon fiber panel under the hood. At left is a Honda racing engine in the Toyota Grand Prix of Long Beach.

Opposite: On May 3, 2004, after a long hiatus, the Ford Motor Company resumed the practice of offering tours of its Rouge plant. Tours of manufacturing plants are just one of the many benefits of being an SAE member.

Aerospace Program Office

The mission of the Aerospace Program Office (APO) is "increasing SAE's impact and value to the aerospace engineering and commercial transport community." In technical areas, its concerns have included deicing/anti-icing, paint/depaint, reliability, cabin automation, training, fasteners standardization, safety, and man-machine interface. The APO has also engaged in cooperative programs with the National Institute for Standards Technology, the Air Transport Association, and the Department of Defense, and it was instrumental in launching A World in Motion II, Challenge 3, the eighth-grade program focusing on aeronautics, specifically the lifting capacity of motorized models.

The PRI Registrar is committed to facilitating its client's achievement of the full benefit intended by the authors of the standards.

The end result of PRI's work has been to improve aerospace supplier quality and to serve as an integral part of the supply-chain management process for many aerospace primes in Europe and North America. The growth of the Performance Review Institute has been strong from the beginning and continues today as more audits are performed and more industry services are provided.

Meetings

A cornerstone of SAE's programs and services, its formal Engineering Meetings structure was established in 1947, but the technical meetings date back almost to the beginning of SAE's existence. In 1906 in New York City, SAE held its first technical meeting, which resulted in the publication of three technical papers and the first volume of the *S.A.E. Transactions*. Since then, SAE has developed and produced meetings in all areas of the mobility communities that it serves.

Long considered to be among the most valuable services SAE offers to its members and the entire mobility community, SAE meetings and symposia account for twenty-five to thirty events each year. Mobility practitioners come together to learn, to present the results of their efforts, and to network with other mobility professionals.

Today, SAE sponsors or administers between fifteen and twenty meetings per year. It also offers about a dozen symposia, which are one- to three-day events focused on emerging technologies, new applications to emerging technologies, or other expanded technical subjects. Additionally, thousands of technical papers are published at SAE meetings annually.

While still holding the successful meetings that have provided value to so many professionals for decades, SAE continues to develop new and exciting meetings to meet the changing needs of today's engineers and managers. For example, its newest meeting is the Commercial Vehicle Engineering Congress and Exhibition (COMVEC), developed for all practitioners in the fields of commercial vehicle and off-highway equipment. Its premier in October 2004 was a resounding success. This event strengthens the products and services SAE provides for this important segment of the mobility community.

Moving ever more rapidly into global realms, SAE is now involved in meetings and seminars around the world. It has also been developing what it calls Technology Theaters for imparting

Stapp Car Crash Conference

Colonel John P. Stapp, an Air Force physician, proved the safety of aircraft ejection seats in daring exploits aboard the jet-powered sled *Sonic Wind*. His experiments aboard a rocket sled at near-supersonic speeds not only probed the upper limits of human tolerance to linear deceleration but also provided information about seat belts and safety harnesses leading to designs that greatly enhanced the survivability of auto crashes. His work also led to formation of the annual Stapp Car Crash Conference, the leading international forum for the presentation of crash-injury research. SAE and the Stapp Foundation have a working relationship where SAE publishes the proceedings of the Stapp Car Crash Conference for the Stapp Foundation. This successful arrangement has been in place for many years.

On September 12, 1955, "Space Surgeon Stapp" made the cover of Time. *Earlier, he had showed off a model of his* Sonic Wind *and told reporters how it felt to accelerate to 421 miles an hour in eight seconds. In the late 1990s, shortly before his death, the colonel was a guest at an NHRA event where top-fuel dragsters were reaching more than three hundred miles an hour in less than five seconds. When he met a driver, the first question he asked was, "Does it feel like getting hit in the back by a locomotive?"*

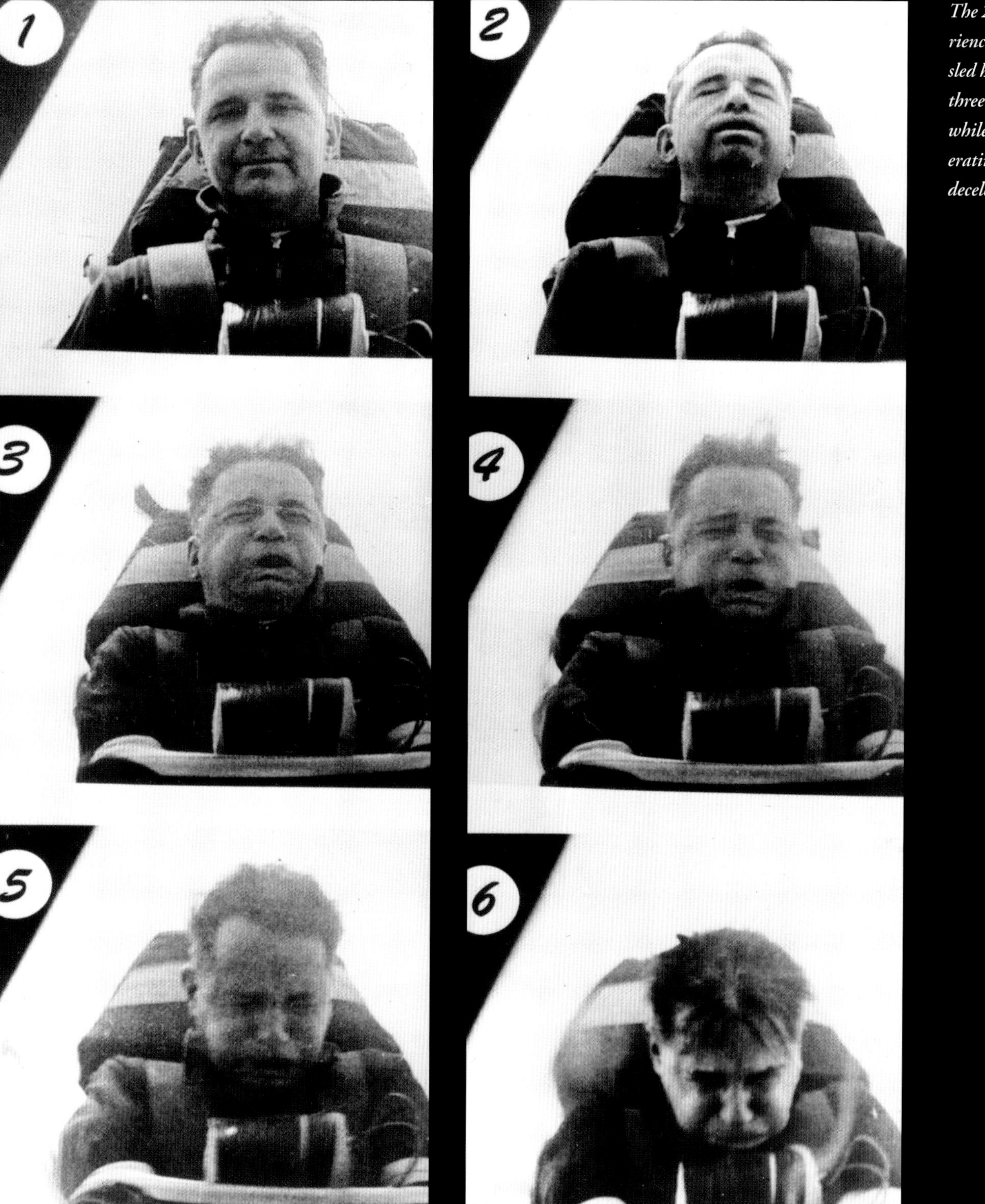

The 22 negative Gfs that Stapp experienced while decelerating after his sled hit its water brake were at least three times the positive Gfs he felt while accelerating. Here, he is accelerating in the first three pictures, decelerating in the last three.

SAE World Congress

The SAE World Congress is the premiere automotive technology event in the world. Held in Detroit, Michigan, for more than seventy years, the World Congress is a combination technical event, management conference, and supplier exhibition.

The World Congress is an international event with delegates representing more than seventy countries and supplier displays from more than twenty countries. Over twelve hundred technical papers are presented during the event each year. Additionally, SAE conducts more than forty professional development seminars and has an onsite bookstore offering a wide range of publications.

Each year a different host company provides organizational leadership and senior OEM and top-tier supplier executives provide guidance through the Industry Leadership Coalition that supports the congress. An executive management conference, featuring more than seventy senior OEM and top-tier supplier executives, is held in conjunction with the World Congress.

The SAE World Congress brings technical experts together and creates a forum for management interaction. It allows companies to showcase their latest products and services, while bringing together engineers, managers, executives, and government and academic practitioners from all over the world together.

cutting-edge information to attendees at its principal convocations. The first of these was presented at the World Congress in 2003, then at COMVEC in 2004, and most recently at a new aerospace event in 2005—thereby reaching out to all three of the Society's core constituencies.

Sections

SAE's local sections provide their membership with something extra: not only technical meetings, but also tours of engineering and manufacturing facilities, opportunities for networking and learning the techniques of public speaking and leadership, and of course, opportunities for community service through educational programs.

Section meetings provide a chance to gain knowledge about cutting-edge technologies. These monthly meetings are arranged by the local governing boards to meet the specific professional and technical needs of their members. Tours of industrial facilities, proving grounds, labs, maintenance areas, and production lines frequently open doors to members that would otherwise be closed to them as individuals.

For fifty-five years Detroit, SAE's largest section, has conducted an annual event now known as the Global Leadership Conference, which attracts automotive leaders from around the world. The Detroit section has also produced the most SAE presidents in modern history. Similarly, the Central Illinois section hosted an annual earthmoving conference for several decades. The first international section outside of North America was the Taipei section. In all, there are more than one hundred local SAE sections all over the world where members meet to advance mobility technology, learn about new developments, and network with their colleagues.

The Blue-Ribbon Panel

In 2000, the SAE board of directors looked at a number of issues facing the society and asked: Where would SAE be in 2010, when it would be well into its second century? To answer this, the board established a blue-ribbon panel led by Don Ableson and made up of people from industry, government, and academia who could contribute fresh perspectives. The panel looked at a great range of issues—the nominating process, broadening the membership base, globalization—and found the most urgent need was for the Society to focus directly on the industries served by its members: aerospace, automotive, and heavy duty. "Wrapping individual membership around industry" was the way executive vice president Ray Morris put it, and he defined the primary imperative as "customizing products and services" to suit the needs of the industries SAE serves.

Perhaps most important, the blue-ribbon panel pointed out that many Society functions could be better managed from Detroit than from Warrendale, and Morris formulated a plan for establishing a much bigger presence in the existing offices in Troy, a Detroit suburb. The amount of SAE space in Troy was tripled, a good part of it being devoted to a learning center that quickly proved to be a huge success, with thousands of people participating in professional development seminars. After the Society embraced the necessity of emphasizing industry sectors, the Detroit office was renamed Automotive Headquarters and Warrendale became World Headquarters.

Sector Vice Presidents

In 2002, the board of directors established three member vice-president positions to assist as SAE moves into the future. These sector vice presidents serve three-year terms and guide SAE policy and programs for the industries in which they serve. Specifically, their responsibilities include:

- advising the SAE president on their respective industry sector
- serving as chairperson for industry program offices
- serving as spokesperson for their industry sector in the absence of the SAE president

These are rotating terms, the first of which was filled for the aerospace sector in 2003. In 2004, a vice president was appointed for the automotive sector, and another one was appointed for the commercial vehicle sector.

The panel also recommended that SAE's board should reflect all professional specialties involved in mobility, including electronics and software engineering, and that SAE's staff organization should reflect the diverse cultures of the world. SAE has now elected its first woman president and its first Asian-born president and is proud of being ahead of most other engineering societies in both respects.

Another crucial accomplishment of the blue-ribbon panel was attaining consensus about the definition of SAE's core competencies as standards and lifelong learning. The focus on standards, of course, goes back almost to the Society's inception, as do meetings and publications. But new avenues of lifelong learning opened with the K–12 program and the Collegiate Design Competitions, as well as professional development seminars. Many of the new seminars every year are customized for specific applications. Industry leaders now understand that if they have a particular educational need, SAE can find a top-flight instructor and develop the curriculum.

The time when firms looked to professional societies such as SAE as organizations that they were obliged to support as part of their community involvement is gone forever. Now, SAE is seen as a supplier just like TRW and Delphi. Industry does not come to the Society for continuing professional development simply because SAE is a professional society. Industry comes to SAE

Because of SAE's excellent reputation, the aerospace industry turns to the Society for audits and also to ensure their aircraft are up to SAE standards. Like most aircraft, this Boeing 737-800 is designed and built in accord with SAE standards and recommendations.

because it has a quality product in the form of seminars and, of course, publications pertinent to design, the area it knows best.

Strategic planning is crucial to any organization that changes officers regularly, and SAE has had an excellent process in place since the early 1990s. In its statement of SAE International Vision, the Society outlined future directions in terms of the question, "What good, for whom?" The statement defines SAE as an international body of mobility engineers who aim to serve society in three ways:

- By fostering the exchange of knowledge through interaction, creativity, timely response to change, social and economic responsibility, and user satisfaction
- By making relevant, timely consensus standards available internationally
- By providing lifelong learning, education, networking, and career development opportunities for mobility engineers and practitioners

SAE has completed a century in motion and is in the midst of a year-long celebration that will include, among other events, products, and celebrations, the 100-Mile Mini Baja. The celebration kicked off in 2004, with historical articles in each issue of *UPdate* on illustrious SAE members such as Charles Kettering and Elmer Sperry.

Looking Ahead

As SAE pauses to celebrate its one hundred years of service to the mobility communities, it is proud of its contributions, not only to the engineering world, but also to society as a whole. SAE has facilitated work that has made automobiles, aircraft, and heavy-duty equipment safer, more efficient, more environmentally friendly, and easier to operate and maintain. From the self-starter in 1911 to today's work on hybrid and alternative-fuel vehicles, the Society has been at the forefront of the technological advances that have helped shape the world.

SAE's role is, and always has been, to assist the engineers and companies that push the technological boundaries of mobility. The Society does not take credit for the advances coming out of corporate research and development departments, university research, government initiatives, or the spark of an idea that the enterprising engineer works to perfect. SAE's role is to facilitate all these accomplishments. SAE bring engineers together. It publishes the standards that are used in so many parts of the mobility industries. It serves as the provider of lifelong learning with programs that begin in elementary school and continue through the working engineer's career. It brings engineers together where their collective energy and creativity can flourish and inspire one another. The spark of ideas can take place at a technical session at one of SAE's meetings, at a panel discussion among experts, or even over a friendly meal between engineers who meet at an SAE event.

This is what SAE has been doing so effectively since 1905. So where does SAE from here? What will the Society be doing in the next few decades? Based on its first one hundred years, one can look ahead to the coming decades and get an idea of what may lie in store.

Engineers will be developing technologies yet undreamed of. SAE will be publishing information about those developments throughout the world. It will be bringing engineers together to discuss technological developments. It will be educating people from a young age throughout their careers, just as it does now.

In whatever form self-propelled mobility takes as the years progress, SAE will be facilitating the work of those professionals to make this technology function better. So it's fair to say that even with the dramatic changes in the ways in which people may move around the world and the ways that information will be managed, the future will find SAE doing what it's done so effectively since 1905. It will be bringing mobility engineers together for the benefit and advancement of mobility technology and of society.

04.01.05
THE BISSELL LIFT-OFF
DIVIDE AND CONQUER.
MEET TV'S FIRST VIDEO GAME FAMILY
GLITCHES AND ALL.
GAME OVER
RETAIL SPACE AVAILABLE
RKF
Contact
Exclusive Agent
212-599-3700
www.rkf.com
LOEWS
THEATRES

Kodak
Share Moments. Share Life.™
share
Kodak
MARQUIS
MARQUIS
ONE WAY
NO TURN
NO TURNS
CHASE

UNITED AIRLINES
Worldwide Service

A Century of Automotive Innovation

Human mobility has made monumental leaps in the past century. Despite many stops and starts, engineers have been able to make men fly and the earth move. The progress of the engineering community has had a profound impact on the modern world, consistently upholding and improving our way of life.

In celebration of engineering's influence, the following timeline features the pivotal automotive inventions, programs, and legislative action over the past one hundred years. Working within both the private and public sectors, the Society of Automotive Engineers has been an integral part of this history, supporting research and creating international forums to translate these developments into books, reports, and industry standards. But this timeline is more than just a chronicle of past achievements—the dedication and hope involved reveal a bright future for the mobility of tomorrow. For as SAE president Charles "Boss" Kettering said, "There exist limitless opportunities in every industry. Where there is an open mind, there will always be a frontier."

1876–1919

1876
Nicolaus Otto debuts what is generally considered to be the first practical four-stroke internal-combustion engine.

1902
Teddy Roosevelt becomes the first U.S. president to ride in an automobile.

1903
At Kitty Hawk, North Carolina, Orville and Wilbur Wright achieve powered, controlled, heavier-than-air flight.

1905
The Society of Automobile Engineers is founded with the aim of promoting "the Arts and Sciences connected with Engineering and the Mechanical construction of automobile vehicles."

1905
The first stolen car is reported, in St. Louis, Missouri.

1906
The first volume of *S.A.E. Transactions* is published.

1906
Alabama sets a maximum speed limit of 8 mph.

1907
Paul Cornu of France is lifted one foot off the ground in a rotary-wing aircraft (helicopter).

1908
General Motors is formed by William C. Durant to manufacture Buicks, Cadillacs, Oldsmobiles, and Oaklands (Pontiacs).

1908
Henry Ford debuts the Model T, of which fifteen million would eventually be produced.

1910
SAE's first insignia is unveiled.

1910
Teddy Roosevelt is the first man ever elected president to fly in an airplane.

1911
Charles Kettering patents the electric starter, first installed on Cadillacs.

Above: The rotary-wing aircraft in which Paul Cornu made the first free flight without the necessity of men on the ground holding it steady.

Left: A distinguished pioneer of mobility, Teddy Roosevelt waves from an automobile during a parade in 1908.

Standard panaceas for urban traffic congestion included the restriction of curbside parking, electric traffic lights at intersection, and the conversion of two-way streets into one-way streets.

1912
The first SAE standard is issued.

1912
The first SAE Exhibit is created.

1913
Henry Ford introduces the first moving assembly line at his Model T plant in Highland Park, Michigan.

1913
The Gulf Refining Company introduces the first drive-in gas station, in Pittsburgh, Pennsylvania.

1914
The first electric traffic light is installed, in Cleveland, Ohio.

1914
Initial publication of SAE's annual compilation of standards in the *SAE Handbook*.

1915
The first hydraulic dumper is used by Parker.

1915
The first SAE Student branch is created, at Cornell University.

1916
The Society of Automobile Engineers, the Society of Tractor Engineers, and the American Society of Aeronautical Engineers merge to create the Society of Automotive Engineers.

1917
After the U.S. enters World War I, SAE engineers design the standardized Liberty airplane engine, tens of thousands of which are manufactured in American automobile factories.

1919
Oregon introduces the first gasoline tax: one cent per gallon, to be used for road construction.

1920–1939

1920
The treasurer of the Bantam Ball Bearing Company of Bantam, Connecticut, Miss Nellie M. Scott becomes the first female member of SAE.

1923
The radio is first offered as a car accessory.

1923
The Lincoln Highway is finally paved in its entirety from New York to San Francisco.

1925
The United States Government develops a numbering system for federal roads.

1925
Fokker debuts the FVIIB-3m, the first modern trimotor aircraft, eventually operated by fifty-four airlines.

1925–1930
Chessie Cummins adapts diesel power, previously restricted to marine applications, to road vehicles.

1926
The first American airplane designed to carry passengers, Ford's Tri-Motor, goes into production; eleven different variants of the "Tin Goose" would be produced before 1932.

1927
The first SAE award is created, the Wright Brothers Medal, recognizing the best paper on the topic of aircraft (and later spacecraft).

1931
Pneumatic tires are introduced for the agriculture industry by Allis-Chalmers.

1931–1935
Diesel engines are adapted to many heavy slow-speed operations.

1933
SAE's Fuels and Lubricants Meetings Committee is formed.

1935
Detergent-based oil is introduced to the automotive market.

1935
The Douglas DC-3 takes flight with a capacity of twenty-one passengers; by 1938 DC-3s would comprise 80 percent of all U.S. airliners.

1935
SAE's Tractor and Industrial Power Equipment Meetings Committee is formed.

1936
SAE's first National Aircraft Production Meeting is held.

1939
The world's first jet engine-powered aircraft, the German Henkel He 178, flies.

The popularity of the radio, first introduced in cars in 1923, soon turned it from an automobile accessory into a mainstream necessity.

Under the command of Richard E. Byrd and Floyd Bennett, this Fokker Trimotor, the Josephine Ford, *was the first airplane to challenge the North Pole.*

1942–1969

1942
SAE's War Activity Council is formed to help coordinate efforts for the Allied forces during World War II.

1943
The Lockheed Constellation takes flight. Equipped with the first pressurized cabin, it is the most advanced plane of its time and ushers in a new era of passenger air travel.

1943
SAE's War Activity Office is established in Detroit (now the Automotive Headquarters).

1944
Howard Hughes and Jack Frye fly a new four-engine, triple-tail Lockheed Constellation from California to New York in less than seven hours.

1944
SAE's Special Publication Department is formed.

1946
SAE's Technical Board begins operation, creating standards used for design, manufacturing, testing, quality control, and procurement.

1947
In a Bell Aircraft X-1A, Chuck Yeager exceeds the speed of sound.

1947
SAE's Engineering Materials Meetings Committee is formed.

1956
President Eisenhower signs the Federal Aid Highway Act of 1956, creating the interstate highway system.

1957
Boeing's 707 becomes the first American-made jetliner to enter commercial service. In the next twenty-one years more than seven hundred 707s would be sold to the world's airlines.

Top: With the help of SAE's War Activity Council, thousands of Jeeps were mass-produced to assist the Allies during World War II.

Above: A Bell X-1A, the aircraft in which Chuck Yeager broke the sound barrier.

The Eagle *has landed. Astronaut "Buzz" Aldrin, Jr., takes a moon walk. Aldrin was joined by Neil Armstrong, the Apollo 11 commander, while co-pilot Michael Collins remained in the command module.*

1958
The Sections Board, approved this year, begins to provide guidance to SAE's network of local sections and advise the board of directors on policy matters relating to section and student activities.

1960
In response to problems caused by the rapid growth in the number of motor vehicles, the California legislature passes the first restrictions on exhaust emissions.

1961
In *Vostok* 1, Yuri Gagarin becomes the first human to orbit the earth.

1961
SAE's thousandth aerospace material specification is published.

1962
Wisconsin introduces the first legislation requiring seat belts as standard equipment in new automobiles.

1962
The child car seat with safety belt is introduced.

1962
John Glenn becomes first American in orbit, ten months after Yuri Gagarin's flight.

1964
The three-engine Boeing 727 introduces passengers to sophisticated high lift devices that make it seem as if the wing is disassembling itself on landing.

1966
SAE Publications add international coverage.

1967
Land, sea, air, and space are added to the SAE logo.

1969
Mack Truck patents cab air suspension.

1969
On July 20, Neil Armstrong and Edwin Aldrin, Jr., walk on the moon.

1970–1990

1970
The first McDonnell Douglas DC-10 takes flight. Designed to operate from both short and long runways, the multi-range jetliner is smaller than the Boeing 747, though its wide cabin gives it a higher passenger capacity.

1970
Airbus Industries formed this year as a multinational effort between Germany, England, and France with a goal of creating a high-capacity twin-jet transport that eventually resulted in the A300.

1971
The Soviet Union launches *Salyut* 1, the first space station. The crew initially slated to man it is unable to enter but a second crew spends twenty-three days in orbit, gaining invaluable spacefaring experience.

1972
The first Lockheed TriStar model, L-1011-1, enters service with Eastern Airlines and Trans World Airlines, powered by Rolls-Royce turbofan engines.

1973
The first commercial passenger car with an air bag, the Oldsmobile Toronado, rolls off the assembly line.

1974
SAE relocates its headquarters to Warrendale, Pennsylvania.

1974
The average family spends 33 percent of its yearly income for a new car.

1974
Federal legislation stipulates a 55-mph speed limit in response to shortages in petroleum products.

1976
On March 6, the British and French inaugurate commercial supersonic service with Concorde.

1976
SAE kicks off its Collegiate Design Series.

1976
The single-seat F-16A Fighting Falcon, a supersonic multi-role fighter jet, flies for the first time.

1977
SAE establishes its Fellow Award in recognition of achievements in technology and engineering.

1978
The SAE Women Engineers Committee is formed.

1978
Japan accounts for more than half the cars imported into the U.S. this year, with total sales passing 1.5 million units.

1980
Electronics are introduced to Off-Highway equipment.

Three thousand Datsuns await customs at a Los Angeles Harbor point of debarkation; by the 1970s more than half the automobiles imported into the United States were Japanese.

A B-2 Spirit Bomber is escorted by two F-117A Nighthawks. The stealth bomber is far more technically advanced than the previous bomber due to its ability to deliver nuclear weapons.

1981
Formula SAE is introduced as an international competition among engineering students for the conception, design, and fabrication of formula-style race cars.

1981
The F-117 Nighthawk, Lockheed's long-awaited "black jet," can evade radar detection while delivering precision attacks on ground targets.

1981
The first volume of *Aerospace Engineering Magazine* is published.

1981
On April 12–14, *Columbia*, the world's first reusable space shuttle, completes its inaugural mission.

1982
The commercial airliner Airbus A-310 flies for the first time.

1983
SAE inaugurates its Professional Development Program, offering continuing education, seminars, and certification to mobility engineers.

1984
The B-1 bomber takes wing. Its primary strength is its capability of a high-speed strike with a large bomb payload for time-sensitive operations.

1985
Case, the agricultural operations company, acquires selected assets of International Harvester's agricultural equipment operations, making Case IH the second-largest farm equipment manufacturer in the industry.

1986
SAE creates a foundation to support educational initiatives, to encourage engineers to become mentors, and to fund scholarships and awards programs.

1989
The B-2 Spirit first takes flight. The stealth bomber represented a major milestone in the U.S bomber modernization program, with the ability to penetrate previously impenetrable defenses and deliver both conventional and nuclear weapons.

1989
SAE's CAESAR Program, the Civilian American and European Surface Anthropometry Research Project, begins this year, digitally scanning and measuring five thousand people for 3-D data on the human body.

1990
SAE establishes the Performance Review Institute (PRI). A not-for-profit organization, PRI develops performance standards and administers quality assurance, accreditation, and certification programs.

1991–2005

1991
SAE develops A World in Motion to foster and enhance the interest of elementary- and middle-school students in mathematics and engineering by means of interdisciplinary curriculum packages.

1991
SAE announces its first international affiliate, SAE Brasil.

1993
SAE opens its Washington, D.C., office.

1993
The first volume of *Off-Highway* magazine is published.

1995
The average U.S. family spends 50 percent of yearly income for a new car.

1996
A World in Motion II, the Design Experience Challenge, integrates the AWIM math and technology curriculum with a curriculum in the language arts.

1996
The SAE Website goes live.

1998
Chrysler merges with Daimler Benz to create Daimler Chrysler.

2001
SAE sponsors the first Environmental Excellence in Transportation awards.

2002
SAE India becomes SAE's second affiliate.

2004
SAE announces its third international affiliate, the United Kingdom.

2004
SAE's membership reaches eighty-four thousand.

2005
The Airbus A-380 is introduced, the largest commercial aircraft ever built.

Joint chairmen Robert Eaton (second from left) and Juergen Schrenepp (second from right) open trading on the New York Stock Exchange. The men are celebrating Daimler Chrysler, the company formed from the merger of Chrysler and Daimler Benz.

A model of the Airbus A-380, the largest commercial aircraft in existence, due for its debut in 2005. The enormous tail of the Airbus hangs in the background.

Pioneers of Industry: The Leaders of SAE

THE PAST ONE HUNDRED YEARS have seen incredible change and advancement in the world. From the invention of the Model T to the creation of the stealth bomber, engineering has kept pace with the progress, at times fueling discoveries that have altered the course of history.

Such growth and innovation requires foresight, cooperation, and leadership. The people on the following pages have embodied these traits while leading SAE from an initial membership of thirty engineers to today's collection of eighty-five thousand engineers, technical professionals, academics, and governmental representatives. Ninety-eight presidents and remarkably, only five chief staff officers, have made crucial decisions that have broadened the scope of influence of the organization, while retaining the original goal of encouraging innovation and cooperation in the automotive world.

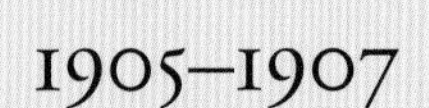

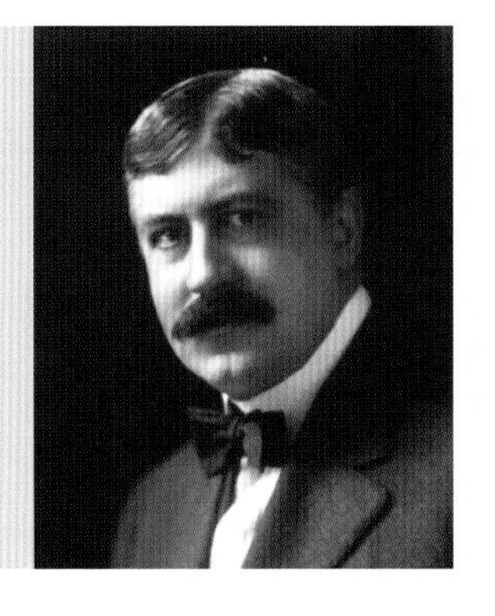

Andrew L. Riker

The Society's first president, Andrew Riker, chief engineer of the Locomobile Company of America, served a three-year term in office. He was elected during 1904, during which the preliminary plans were perfected for the founding of the Society in 1905. He was active in all aspects of automobile design, having developed gasoline, steam, and electric vehicles.

1908

Thomas J. Fay

Thomas Fay, electrical engineer and consultant, was president in 1908. He was a generous contributor of significant technical papers and discussions at the Society's first two annual meetings in 1906 and 1907, lending high professional tone to such subjects as materials, standards, vehicle durability, and the performance of professional duties.

1909

Henry Hess

SAE's 1909 president was Henry Hess, of Hess-Bright Manufacturing Company of Philadelphia. He was a frequent contributor to technical and policy discussions, and was an expert on bearings and their applications. He foresaw the growing managerial needs of the organization and undertook initial talks with Coker Clarkson, seeking to convince him to become the Society's first chief staff officer.

1910

Howard E. Coffin

The Society's fourth president was Howard Coffin, a highly accomplished engineer and executive who was then vice president of engineering for Hudson. His initiative and fundraising ability provided for the first secretary and general manager of the Society. He also stimulated a vigorous standards program, laid groundwork for the *SAE Bulletin*, and put SAE on a sound fiscal basis.

Henry Souther

SAE's fifth president was Henry Souther, educated at MIT and in Europe. His knowledge of chemistry, steelmaking, and engineering led him to become technical consultant to the state of Connecticut and to the Locomobile Company. He was the first chairman of the SAE Standards Committee and helped establish Langley Field, the army Air Corps, and army motor-vehicle transportation.

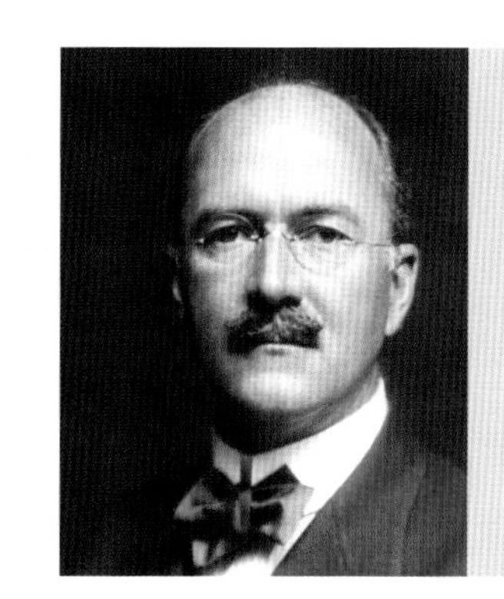

1911

Henry L. Donaldson

When he became SAE president in 1912, Henry Donaldson was an automotive engineering consultant with a background as a writer and editor. He died during the summer of 1912. Early in his term, however, the *SAE Bulletin* was enlarged, peer review of technical papers was initiated, and *S.A.E. Transactions* was copyrighted to prevent misuse.

1912

Herbert W. Alden

Herbert Alden completed Henry Donaldson's 1912 term and was later elected president in his own right in 1923. As an engineer, he joined the Pope Manufacturing Company in 1895, ultimately heading the Timken-Detroit Axle Company. He was awarded the Distinguished Service Medal for ameliorating tank design problems during World War I.

1912

Howard Marmon

The Society's eighth president, Indianapolis engineer Howard Marmon, began his career as a builder of grain mill equipment but soon entered the automotive world and became builder of the Marmon automobile. He also built the Marmon Wasp, winner of the 1911 Indianapolis 500-mile race. He initiated the Research Division of the SAE Standards Committee.

1913

Henry M. Leland

Henry Leland, founder of Cadillac Motor Car Company and winner of the Dewar Trophy in England for interchangeability of auto parts, was the Society's 1914 president. He did much to spur development of sections and student memberships, and explored the possibility of a system of primary elections for SAE officers.

1915

William H. Van Dervoort

William Van Dervoort, 1915 president, came from Moline, Illinois, where he headed an engineering firm and the Moline Auto Company. During his term, the Society's activities expanded into tractor, marine, industrial, and aeronautical areas. In his presidential address he exclaimed, "With what force has the motor car entered the field of transportation!"

Russell Huff

Russell Huff, chief engineer of Dodge Brothers Company, served as president in 1916. He led the activities of the Society in aiding the national defense program, presided while arrangements were made for the merger of aeronautical, tractor, and marine groups with SAE, and saw the start of substantial growth of membership and technical activities.

George W. Dunham

George Dunham, chief engineer of Chalmers Automobile Company, led the Society through its first year of active participation with the U.S. Government in time of war. During his term, the name *automotive* was first applied to the Society, and its periodical publication was re-named the *SAE Journal*.

Charles F. Kettering

One of the most famous automotive engineers in history, Charles Kettering, was SAE president in 1918. Widely known for his invention of the self-starter, he was active in many areas and became director of General Motors Research. He urged cooperation between the petroleum and automotive industries, which led to formation of what is now the Coordinating Research Council.

1918

Charles M. Manly

The fourteenth SAE president was Charles Manly, known for his work in collaboration with the noted aeronautical researcher Professor S. P. Langley. He was a consultant to the U.S. and French governments in aeronautical matters, and, while president, stressed sections work and development of industrial understanding of standards.

1919

Jesse G. Vincent

Jesse Vincent, 1920 president, played a major part in development of the Liberty aircraft engine during World War 1. He started his engineering career with Burroughs Adding Machine Company and then worked at Hudson and at Packard. During his term, SAE membership passed five thousand for the first time.

1920

David Beecroft

David Beecroft, elected president for 1921, was an academic professor in his native Canada before becoming technical writer and editor of several prominent automotive magazines. In addition to serving as president, he was also SAE's treasurer from 1933 to 1943.

1921

Benjamin B. Bachman

B. B. Bachman, 1922 SAE president, worked on passenger car and truck development for many years with Autocar Company of Ardmore, Pennsylvania. He was active in the transition from solid rubber truck tires to pneumatic tires. As SAE's treasurer from 1944 to 1959, Bachman brought wisdom and fiscal guidance to the Society's leadership for many years.

Herbery W. Alden

President of SAE for part of 1912, Herbert Alden was elected president in his own right in 1923. The engineer began his career with the Pope Manufacturing Company in 1895 and retired as head of the Timken-Detroit Axle Company. He was awarded the Distinguished Service Medal for ameliorating tank design problems during World War I.

Henry M. Crane

The varied career of 1924 SAE president Henry Crane included work for AT&T and Western Electric Company. He founded Crane and Whitman Company to develop gasoline engines, which became Crane Motor Car Company and eventually Simplex Automobile Company. In 1920, via other mergers, Crane became vice president of engineering of Wright Aeronautical Company.

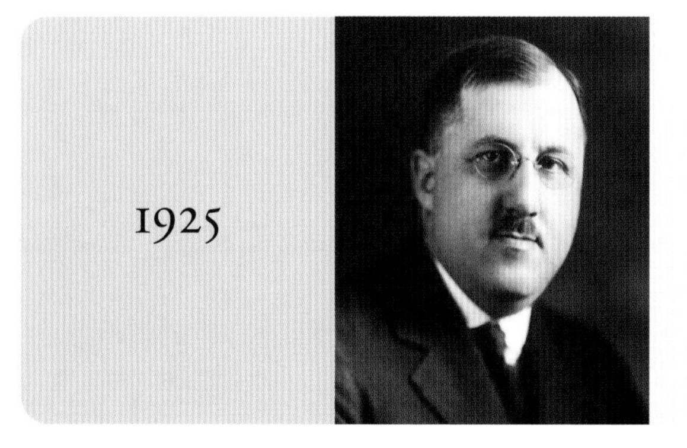

Harry L. Horning

Harry Horning was first noted as designer of the structure and operating mechanism of the Duluth Steel Bridge. In 1906 he founded Waukesha Motor Company. Active in the Society of Tractor Engineers and the National Gas Engine Association, he later helped them merge into SAE. As 1925 president, he helped lead the Society in its development of cooperative fuels research.

Thomas J. Litle, Jr.

Thomas Litle, Jr., 1926 president, had successful careers in the fields of automobiles, illumination, and refrigeration. He joined Cadillac in 1917 and Lincoln in 1918 in research and experimental engineering, and became chief engineer of Marmon in 1926. While he was president, the *SAE Handbook* was first issued as a bound volume.

1926

John H. Hunt

John Hunt, a University of Michigan electrical engineering graduate, joined Packard in 1912, Dayton Engineering Laboratories in 1913, and became head of the electrical division of General Motors Research in 1920. He later headed GM's New Devices Section. As SAE president in 1927, he contributed much to the development of policies on research.

1927

Wiliam G. Wall

William Wall, 1928 president, was chief engineer of National Motor Vehicle Company, Indianapolis, when he introduced what was described as America's first six-cylinder car. He also built a winning racer for the Indianapolis 500. He was active in several unusual engineering projects, such as the changeover of submarines from steam to internal combustion power.

1928

William R. Strickland

SAE's 1929 president, William Strickland, began his career at American Radiator Company and then joined Peerless Motor Car Company in 1913. In 1922 he began work at AM Research, ultimately becoming the assistant chief engineer of Cadillac. During his term, the Society was organized into professional activity groups, each under a vice president.

1929

Edward P. Warner

Twenty-fourth SAE president Edward Warner was Assistant Secretary of the Navy for Aeronautics when he was elected in 1930. He had been an engineer for the military air service, chief physicist for the National Advisory Committee for Aeronautics, and widely known as an aeronautical engineering professor at MIT.

Vincent Bendix

Vincent Bendix, the 1931 SAE president, was founder of Bendix Aviation Corporation and other Bendix enterprises. He gave personal funds for SAE research in riding comfort and strongly supported SAE's technical program. During his term, a careful study was made of student branches, and operational cost reductions were implemented to cope with the effects of the Depression.

Arthur J. Scaife

Arthur Scaife began his career at White Sewing Machine Company's automobile department and remained there for most of the rest of his life, designing White motor vehicles. His prudent administration as 1932 SAE president offset much of the depression's economic effect. Research on diesel fuels and lubricants was increased during his term.

Hobart H. Dickinson

Hobart "Doc" Dickinson, the 1933 SAE president, spent his entire professional career at the National Bureau of Standards. During his term he made a nationwide automobile tour visiting SAE sections. An authority on thermodynamics and internal-combustion engines, his interests also included headlighting, engineering management, riding comfort research, and traffic safety.

Delmar G. Roos

1934

Delmar "Barney" Roos, who served the Society as its 1934 president, worked on radio and turbine development after his 1911 graduation from Cornell. After work at Locomobile and Marmon, he joined Studebaker in 1926 and became its chief engineer. After working as consultant to the Rootes Group in England, he joined Willys and is noted as the father of the World War II Jeep.

William B. Stout

1935

William Stout, SAE's 1935 president, had a career that included activities as engineer, inventor, writer, and speaker. Among other things, he designed the Imp cyclecar in 1910 and worked for United Aircraft Engineering, where his most notable achievement was his development of an all-metal cantilever monoplane shortly after World War I.

Ralph R. Teetor

1936

Ralph Teetor, 1936 president, graduated from the University of Pennsylvania in 1912 and eventually became president of Perfect Circle Piston Ring Company. He invented the Speedostat, the cruise-control device used on many automobiles. During his term, the first National Aircraft Production Meeting was held. He established the Teetor Educational Awards Program in 1964.

Harry T. Woolson

1937

Harry Woolson, executive engineer at Chrysler while 1937 SAE president, was initially a marine engineer. After he worked with companies building gas engines for boats, he then switched to land vehicles as truck engineer and subsequently worked at Packard, Studebaker, Maxwell, and other firms. As SAE president, he encouraged younger engineers to become active in the Society.

1938

Clarence W. Spicer

Clarence Spicer, 1938 SAE president and 1904 Cornell graduate, founded Spicer Universal Joint Company in 1905. During World War I he was involved in the design of the Class B truck. As SAE president he laid the foundation for more extensive activities in, and cooperation with, railway engineering, helping to point out the many common interests of this field with automotive engineering.

1939

William J. Davidson

SAE's thirty-third president, William Davidson, was born in Montreal. After graduation from McGill University, he served in France with the Motor Transport Corps and was awarded the Cross of the Legion of Honor by the French government. During his 1939 term as president, he stimulated SAE's relationships with foreign engineers and led an SAE World Engineering Congress.

1940

Arthur Nutt

1940 SAE president Arthur Nutt graduated from Worcester Polytechnic Institute in 1916 and spent most of his professional career with Wright Aeronautical Corporation and its predecessors as an authority on air-cooled and water-cooled aircraft engines. As president, he played a major part in stimulating SAE aeronautical standards work to render major service to industry.

1941

Archie T. Colwell

Arch Colwell, graduate of the U.S. Military Academy and the Army Engineers School, spent four years in the service before joining the Steel Products Company. He served SAE as president in 1941 but continued an active and vital role for many years after, chairing the SAE Finance Committee for twenty-five years and achieving industry funding for the Cooperative Engineering Program.

Arthur W. Herrington

Arthur Herrington dedicated his engineering career to the development of better off-road military and civilian vehicles. He became president of Marmon-Herrington Company in 1931 and was elected to the SAE presidency in 1942. During his term, the SAE Automotive Technical Advisory Board became the War Engineering Board and greatly expanded its scope of action.

1942

Mac Short

Mac Short, World War I pilot and 1943 SAE president, spent his engineering career in aviation. One of the founders of Stearman Aircraft Company, he became its vice president of engineering, and later a founder and vice president of Lockheed Aircraft Company. His term of office saw expanded results from SAE's aeronautical and technical committees in the war effort.

1943

William S. James

William James started his career at the National Bureau of Standards. When he was elected to the Society's presidency in 1944, he was chief engineer of Studebaker Corporation, and later became vice president of engineering and research with Fram Corporation. He saw SAE's World War II work near its completion and established its Special Publications Department.

1944

James M. Crawford

James Crawford, 1945 president, began his career with the American Motors Company in 1907. He became vice president of engineering for General Motors, and led the beginning of SAE's postwar readjustment to member and industry needs via reorganization of the Coordinating Research Council and plans for the establishment of the Technical Board.

1945

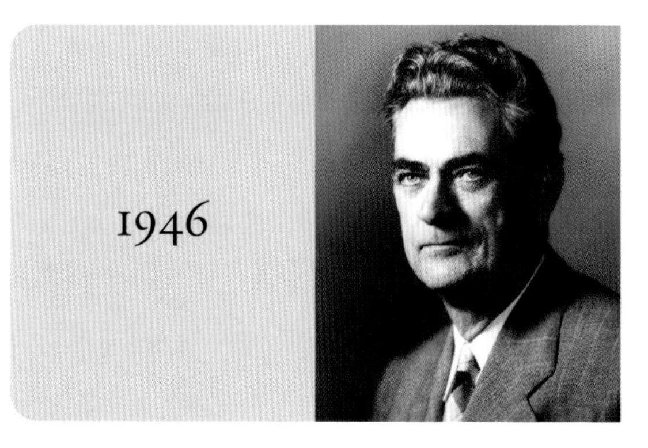

L. Ray Buckendale

Ray Buckendale, the Society's 1946 president, graduated from the University of Michigan and began a life-long career with the Timken-Detroit Axle Company. He served SAE in many posts, and during his term of office saw the SAE Technical Board swing into full operation to guide and direct all SAE technical committee activity.

C. Erwin Frudden

The first tractor engineer to serve as SAE president was Erwin Frudden, elected in 1947. He became part of the tractor industry in 1910, worked in tractor plants, and became chief engineer of Allis-Chalmers Manufacturing Company in 1938. He stabilized the Society's finances by aiming for reserves equal to one year's operating expenses.

Reginald J. S. Pigott

Reginald Pigott was chief engineer of Gulf Research and Development Company while SAE's 1948 president. His career included designing steam powerplants, metal manufacturing, industrial plant design, professorship at Columbia University, and writing forty technical papers on various subjects. He visited and talked to every one of the Society's sections and groups during his term.

Stanwood W. Sparrow

Stanwood Sparrow, 1949 SAE president, was also vice president of engineering at Studebaker. He was nationally known as a top engine and research expert and was a prolific author of SAE papers, including one milestone titled, "My Friend the Engine." He emphasized service to SAE and once said, "Whatever is to be received from the Society, we the members must put into it."

James C. Zeder

James Zeder, SAE president in 1950, was vice president of engineering of Chrysler Corporation. His outstanding contributions to SAE spanned many years, including service on the War Engineering Board and chairmanship of the Technical Board. His term of office saw membership pass fifteen thousand.

1950

Dale Roeder

Dale Roeder, the 1951 SAE president, spent his career with Ford Motor Company. After beginning as a draftsman, he moved up to become the executive engineer for commercial vehicles. His responsibilities at Ford included tractors, trucks, and operation of twenty-two hundred acres of Ford's experimental farms. He also participated in projects to subject tractors to stress analysis.

1951

Daniel P. Barnard, IV

Fuels and lubricants specialist Daniel Barnard was the Society's forty-sixth president in 1952. After earning a doctorate in chemical engineering, he joined Standard Oil of Indiana in 1925 and later established its automotive laboratory. Most of his career was spent "fitting fuels to engines," and he received the Horning Memorial Medal in 1944 for distinguished service in this area.

1952

Robert Cass

Robert Cass became the fourth English-born SAE president in 1953. After service in the Royal Air Force as an aeronautical engineer, he came to the United States to teach at Harvard, then spent over two years in Japan teaching construction of flying boats. Later, he joined White Truck Company as chief engineer, then became assistant to the president of that firm.

1953

William Littlewood

William Littlewood, the 1954 SAE president, had a career in air transportation notable for the design of the Douglas aircraft family from the DC-3 through the DC-7. Joining the predecessor of American Airlines, he instituted a program in which operators of transport aircraft helped specify the type of craft they purchased. In 1935 he won the SAE Wright Brothers Medal.

C. G. Arthur Rosen

Arthur Rosen became SAE president in 1955, the Society's fiftieth anniversary year. Active in diesel design until 1923, he taught at the University of California until he joined Caterpillar Tractor Company in 1928 to develop tractor diesel engines. During World War II he worked on several major projects relating to submarines, including sustained underwater operations and noise detection.

George A. Delaney

George Delaney, the 1956 SAE president, worked at Savage Arms Company, Paige-Detroit, Graham-Paige, and Pontiac. After retiring from Pontiac as chief engineer, he served as consultant and special project manager of the emissions control research program of the Automobile Manufacturers Association. He was both a local and national SAE treasurer for many years.

W. Paul Eddy

When Paul Eddy was elected president in 1957, he was chief of engineering operations at Pratt and Whitney Division. He contributed greatly to SAE, both in technical and administrative areas, and urged during his term of office that engineering education be broadened to include a wider variety of cultural subjects.

William K. Creson

1958

William Creson, 1958 president, spent his working life with Ross Gear and Tool Company, completing his career as its vice president of engineering. He was a widely known authority on the application of power steering to wheeled vehicles of all types. He served SAE in many capacities. During his term of office, he helped to accelerate the SAE Planning for Progress reorganization.

Leonard Raymond

Leonard Raymond served as SAE's 1959 president. He started his career in the petroleum industry with the Texas Company in 1928, eventually joining Socony Mobil and became their chief automotive engineer for research. His SAE work included valuable contributions to the Planning for Progress reorganization, key technical papers, and service as SAE's international ambassador of good will.

Harry E. Chesebrough

1960

Harry Chesebrough, SAE's 1960 president, spent his career at Chrysler Corporation in test work, engineering, and management. He served for two years as president of the American Standards Association, while changing it to the American National Standards Institute. He chaired the SAE Planning for Progress program and helped carry out its plans during his 1960 term.

Andrew A. Kucher

1961

The holder of nearly one hundred patents, Andrew Kucher, the 1961 SAE president, enjoyed careers in many fields. In 1951 he formed the Ford Motor Company Scientific Laboratory, and was later named vice president, engineering and research, for Ford, where he managed one of the finest industrial engineering and research centers in the world.

Frank W. Fink

Frank Fink, the SAE president in 1962, spent his career in the aircraft industry. After working at Curtiss-Wright and Convair, he joined Ryan Aeronautical Company, becoming its vice president of engineering and products. He held many key posts within SAE and contributed to the cross-fertilization of aircraft and ground-vehicle engineering disciplines.

Milton J. Kittler

After graduation from Illinois Institute of Technology, Milton Kittler, the 1963 SAE president worked as an engineer with International Harvester, then with the Stromberg Carburetor Division of Bendix Aviation. He spent most of his career with Holley Carburetor Company from which he retired as president. Active on many SAE committees, he was SAE national treasurer for ten years.

John T. Dyment

The first Canadian resident to serve as SAE president was John Dyment, elected in 1964. As chief engineer of Air Canada, his career was spent improving transport aircraft, in which he pioneered many developments relating to low-temperature operation. He served in many SAE capacities and was noted for his advancement of the study of human factors involved in aircraft flight.

John R. MacGregor

John MacGregor, 1965 president, built a bicycle-mounted glider when he was twelve, and built his own automobile before he was twenty. His professional career was spent in the petroleum industry; when he became SAE president he was the director of the California Research Corporation, the research arm of Standard Oil of California.

George Edwin Burks

1966

Edwin Burks, 1966 SAE president, spent most of his professional career at Caterpillar Tractor Company where he was responsible for major increases in diesel engine output. He joined the company in 1929, became its chief engineer in 1942, and its vice president of engineering and research in 1955. He served SAE in nearly every possible capacity before becoming its president.

Ralph H. Isbrandt

Ralph Isbrandt started work at A. O. Smith Company as a chassis detailer, and then moved to Firestone as a research engineer. After the war, he worked at Kaiser-Frazer, eventually becoming vice president of engineering for American Motors. The Ralph H. Isbrandt Automotive Safety Award was established by SAE in recognition of his contributions to automotive engineering and safety.

F. Burrows Esty

Bill Esty, SAE president in 1968, began his career working on aircraft magnetos at Bendix Scintilla. In 1948 he moved to Wisconsin Motor and rose to vice president of engineering in 1957. He played an important role in the development of the first Wisconsin overhead-valve engine, and was intimately involved with the standardized parts program and the manufacture of military engines.

Phillip S. Myers

1969

Phillip Myers, SAE's first academic president, was elected in 1969. Known for his research into heat transfer and combustion, he is the author or co-author of more than fifty technical publications. He has earned several SAE awards, including the Horning Memorial Award and the Arch T. Colwell Award. He served SAE in many areas before and after his term as president.

Harry F. Barr

In 1970 the Society had as its president Harry Barr, a top engineering executive at General Motors. Beginning at Cadillac in 1929, he moved to Chevrolet in 1952, became its chief engineer in 1956 and then headed the entire GM engineering staff in 1963. He was instrumental in the development of the overhead-valve family of GM engines, as well as GM's automatic transmission line.

Harold L. Brock

Harold Brock, the 1971 president, began his career at Ford Motor Company where he became acquainted with Henry Ford, Sr., and was one of only seven engineers retained after the 1933 bank crash. In 1959 he moved to John Deere Company as assistant director of engineering research, and was made manager of vehicle engineering, responsible for agricultural tractors throughout the world.

Eugene J. Manganiello

Gene Manganiello, SAE's 1972 president, has spent much of his career at National Aeronautics and Space Administration, where he started at Langley Research Center as head of its heat transfer section, and rose to the position of its deputy director. He is the author of some thirty-five technical reports and papers, and has held more than fifteen top-level posts within SAE.

John C. Ellis

Automotive emissions specialist Jack Ellis was elected SAE president in 1973. After World War II, he joined Shell Oil Company and received research assignments in various parts of the world. At the time of his election, he was manager of vehicle emissions control and technical information, with responsibility for Shell's corporate-wide efforts to reduce vehicle emissions.

Wilson A. Gebhardt

Bill Gebhardt, engineer and lawyer, was the 1974 SAE president. After teaching at Case School of Applied Science, he moved to Bendix Corporation. He held a variety of positions and was executive assistant to the general manager of Bendix Engine Controls Division at the time of his election. During his term, SAE moved its headquarters from New York to Warrendale, Pennsylvania.

1974

George J. Huebner, Jr.

George Huebner, known as the father of the automotive gas turbine, was SAE's 1975 president. His forty-three-year career with Chrysler Corporation included twenty years of research and resulted in major gas turbine advances. He was also one of the first to foresee the energy crisis of the 1970s. He contributed for many years as a director of the Coordinating Research Council.

1975

Rodger Ringham

Rodger Ringham, 1976 SAE president, was trained as an aeronautical engineer. In 1969 he turned to land vehicles, joining International Harvester to be responsible for on-highway and off-highway vehicles and turbines. He has been productive in many areas of SAE work, particularly in SAE's relationships with industry, government, and sister organizations in standardization.

1976

Gordon L. Scofield

The Society's second academic president, Gordon Scofield, was elected for the 1977 term. A specialist in radiative heat transfer and the combustion process, his contribution to the automotive industry was generating engineering graduates. He has been active in establishing liaisons between academe and industry, to keep the academic community abreast of where industry is heading.

1977

Leo A. McReynolds

Petroleum specialist Leo McReynolds served the Society as its 1978 president. During his long career with Phillips Petroleum Company, he held a variety of positions culminating in manager of the Environment and Consumer Protection Division. While SAE president, he worked for greater awareness of energy conservation on the part of the engineering community and the general public.

Lewis E. Fleuelling

SAE's 1979 president, Lewis Fleuelling, was employed by Monroe Auto Equipment Company, ultimately becoming senior vice president. As SAE's president, he emphasized the need for improved communications and relationships between government and industry. His chairmanship of the SAE Finance Committee continued the emphasis on the Society's fiscal integrity.

Harold C. MacDonald

Harold MacDonald was the SAE president for SAE's seventy-fifth anniversary year. After employment at Packard and General Motors, he joined Ford Motor Company in 1948 and rose to his position of vice president of the engineering and research staff. His SAE work encompassed virtually every area within the Society's structure.

Philip J. Mazziotti

SAE's 1981 president, Philip Mazziotti, has been a member of SAE since 1949. Since that time, he was chairman of the Engineering Activity Board, the Technical Board, and also the Finance Committee. He worked at the Dana Corporation beginning in 1941 and became their vice president of product. In 1981 SAE membership passed forty thousand while under Mazziotti's leadership.

N. John Beck

N. John Beck, SAE's 1982 president, was a World War II naval officer. After working for Douglas Aircraft Company, Cummins Engine Company, and Rohr Corporation, he founded his own engineering and management-consulting firm, BKM, Inc., in 1975. He has been involved in SAE since 1952 and is remembered for his strong encouragement of activities at the local level.

1982

Charles C. Colyer

1983 SAE president Charles Colyer worked in lubricant research since 1947, and retired as the assistant to the senior vice president of research at Lubrizol Corporation. A member of SAE since 1950, Colyer has been the chairman of the Fuels and Lubricants Technical Committee, and is remembered for spear-heading the formation of the Lubricant Review Institute in SAE.

1983

Gordon H. Millar

Gordon Millar, SAE's president in 1984, worked with Deere and Company since 1963 and retired as their vice president of engineering. A former member of the SAE Board of Directors, Millar is also an SAE Fellow. He holds seven patents and is the author of many technical papers. He is remembered at SAE for accelerating the creation of the Engineering Education Board while president of SAE.

1984

Elliott A. Green

SAE's 1985 president Elliott Green was with the Lockheed-California Company and retired as their vice president and general manager of Product Support. He has held a number of SAE positions, including chair of the Aerospace Council. During his presidency, he brought attention to the aero-space community within SAE.

1985

Franklin Walter

Franklin Walter joined Chrysler Corporation in 1948 and retired as their director of Corporate Timing Planning for Product Development. He has been a member of SAE since 1951 and is remembered for his interest in local sections, visiting more than fifty existing and potential sections during his 1986 term as SAE president.

William S. Coleman

William Coleman was the 1987 SAE president. He retired as the director of Eaton Corporation's New Product Opportunities for the Corporate R & D Detroit Center. Active in SAE since he became a Junior Activity member in the 1950s, he served as a member of the board of directors and as the Detroit section chairman, and was a generous contributor of technical papers.

George D. Aravosis

SAE's 1988 president was George Aravosis. He retired as the director of Technology Planning for Navistar International Transportation Corporation. He has been extremely active in local and national levels of SAE: in 1983 he was on five different committees simultaneously, chairing two of them. While president, he led SAE in hosting a FISITA conference.

Edward T. Mabley

Ed Mabley, the SAE president in 1989, helped design the first automatic truck transmission while at GM. He moved to Ford in 1965, and retired as Ford's Heavy Truck Product Development manager. He has been a member of SAE since 1958. A cofounder of SAE's Truck and Bus Meeting, he also played a major part in initiating a fundraising program for engineering education support.

John L. Mason

1990

1990 SAE president John Mason joined the Garrett Institute in 1950, and became the VP of Engineering and Technology for Allied-Signal Aerospace, Garrett's successor company. He has been on advisory boards for NASA and the National Institute of Standards, and at SAE he has been on several national boards, and is remembered for his role in establishing the SAE Foundation.

Lamont Eltinge

1991

Lamont Eltinge, SAE's 1991 president, worked in the fuels industry for many years and was the director of research at the Eaton Corporation. From 1989 to 1990, he was one of two SAE Fellows in the White House Office of Science and Technology Policy and was also the recipient of an SAE Horning Medal Award.

Jack W. Schmidt

1992

Jack Schmidt started with General Motors after graduating from General Motors Institute in 1954. He retired as the company's director of Powertrain Systems. Active in SAE's Indiana and Detroit sections, he has also chaired the VISION 2000 Program Office, and did much to inaugurate a new standardized vision development process. He holds five patents.

Bruce R. Aubin

1993

A leader in the air transport industry, Bruce Aubin was SAE's 1993 president. He retired as the senior vice president of Maintenance Operations for USAir. He is an SAE Fellow, and has worked diligently to guide the board of directors in pursuing two strategic initiatives: Environment/Total Life Cycle Technology and Globalization. He is also a multi-engine licensed pilot.

Randall R. Richards

The 1994 SAE president, Randall Richards, began his engineering career working for Caterpillar in 1970 and is now its manager of corporate sustainability. He has been active in SAE since 1980. As both a member of the Brazil Advisory Committee and the Development Program Coordinating Committee, he is best remembered for his work developing and supporting the SAE Brasil affiliate.

John M. Leinonen

John Leinonen's career at the Ford Motor Co. spanned thirty-five years in the Truck and Auto Safety areas. He retired while president of SAE to become Group VP and Principal at Exponent Inc., retiring there in 2001. He has sat on or chaired many SAE boards and committees. He continues to be one of SAE's most active past presidents and co-chairs SAE's 100th Anniversary Committee.

Claude A. Verbal

Claude Verbal joined Buick Motor Division in 1964. He worked in many different General Motors divisions before retiring as the plant manager in Service Parts Operations for GM. He has been an active member of SAE for thirty years and is remembered for his leadership in identifying strategies to achieve SAE's total life cycle end in manufacturing.

David Holloway

1997 SAE president David Holloway began teaching at the University of Maryland in 1971. He formed the SAE student chapter there in 1978 and remained the faculty advisory until his retirement. He and his students won their first Mini Baja in 1982. He has remained focused on alternative fuels and is best remembered in SAE for his support of engineering education.

Ronald K. Leonard

1998 SAE president Ronald Leonard began his career with John Deere. He retired in 1998 as director, Worldwide Agricultural Tractor & Component Engineering. He has been active in SAE for many years, and among other responsibilities he was the general chairman of the International Off-Highway and Powerplant Congress and Exposition in 1997.

1998

Donald W. Ableson

Donald Ableson, the 1999 SAE president, worked at General Motors. He retired as the director of Special Vehicle Activity with the North American Car Group. Active in SAE for more than thirty-five years, he held several positions, including a member of the board of directors, chairman of the Detroit section, and president of the SAE Foundation Canada Board of Trustees.

1999

Rodica A. Baranescu

Rodica Baranescu was the first female president of SAE. She obtained her doctorate in Romania before coming to the United States in 1980. A winner of the Forest McFarland Award and an SAE Fellow since 1999, she is remembered for advancing international relationships, including helping to establish two Romanian Joint SIAR-SAE Membership Groups.

2000

Neil A. Schilke

Neil Schilke worked in several positions over forty years at General Motors. He retired as General Director of Engineering. He was a founding director of SAE Foundation Canada and co-founder of the SAE Automotive Resources Institute. His SAE awards include the 2001 Medal of Honor, naming the SAE Foundation Cup in his honor, and election as an SAE Fellow.

2001

S. M. Shahed

S. M. Shahed is vice president of Advanced Products and Systems at Garrett Engine Boosting Systems. He has served on many SAE boards, has won many awards for his technical papers, and has taught at the University of California, Berkeley, and the University of Texas, San Antonio. He is remembered at SAE for selecting and establishing the first sector vice president.

Jack E. Thompson

2003 SAE president Jack Thompson has spent his entire career with Chrysler, designing the first minivan and founding Chrysler's Engineering Book of Knowledge. He retired as the director for computer-aided engineering and concept development for all seven Platform Teams. He has served on the SAE Foundation Board of Trustees, and was selected as an SAE Fellow in 1999.

Duane Tiede

Duane Tiede was SAE's 2004 president. He was with Deere and Company for twenty-one years before moving to GNH Global and retiring as their vice president of functional engineering. Active at both SAE's national and section levels, he has served on the board of directors, the Off-Highway Continuity/Advisory Committee, and has written several technical papers.

J. E. "Ted" Robertson

Serving as president for SAE's centennial year, Ted Robertson began his career with General Motors and after thirty-four years there he joined ASC as its vice chairman for Product Development. He has served SAE in almost every capacity, from president of the student branch at the University of Toronto to a member of the blue-ribbon panel, which helped determine the future of SAE.

Coker F. Clarkson

SAE's first Secretary & General Manager, Coker Clarkson served from 1910 to 1930. He had been the secretary of the mechanical branch and assistant general manager of the Association of Licensed Automobile Manufacturers. He oversaw SAE's transition into a mobility organization that included aircraft and heavy-duty vehicles.

1910–1930
CHIEF STAFF OFFICER

John A. C. Warner

SAE's Secretary & General Manager from 1930 until 1960, John Warner was a mechanical engineer. He led SAE through World War II and helped to establish the War Engineering Board and its committees, which led to the SAE Technical Board structure. After the war, he led SAE into adapting to the growing needs of a population hungry for new transportation machines and technology.

1930–1960
CHIEF STAFF OFFICER

Joseph Gilbert

Becoming only the third person to hold his post, Joe Gilbert was SAE's Executive Vice President from 1960 until 1985. Among his many accomplishments, Joe is remembered as the leader who moved SAE from New York to Warrendale. This is credited with helping SAE maintain a competitive cost structure and providing the foundation for the growth that SAE has experienced since 1974.

1960–1985
CHIEF STAFF OFFICER

Max E. Rumbaugh, Jr.

Appointed to the position of Executive Vice President in 1986, Max Rumbaugh served through 2001. One of Max's major accomplishments was to focus SAE on growing as a global organization. He implemented the policy of promoting international membership, sections, meetings, and affiliates. He also helped SAE focus on emerging mobility technologies earlier in their developmental phase.

1986–2001
CHIEF STAFF OFFICER

Raymond A. Morris

After working for SAE for twenty-eight years, Ray Morris was appointed Executive Vice President and Chief Operating Officer in 2002. He successfully reconfigured the Detroit Branch into the SAE Automotive Headquarters to better serve members in the automobile industry. He is continuing global SAE's development and emphasizes providing educational programs for students and increasing benefits to younger members.

2001–
CHIEF STAFF OFFICER

Index

Photo Credits

AFP/Getty Images: 25

©Nogues Ala/CORBIS SYGMA: 135

American Stock/Getty Images: 45, 50

AP/Wide World Photos: 8–9, 10, 18, 49a, 49b, 95b, 108, 109, 111, 112, 113, 116, 118, 122, 125, 132, 133a, 133b, 134, 139a, 141, 142b, 158

AvroArrow.org: 92

©Bettmann/CORBIS: 21, 23, 28a, 46, 52–53, 55, 76b, 77, 81, 82b, 83, 91, 94a, 95a, 154, 162–163

Torsten Blackwood/AFP/Getty Images: 138a

Margaret Bourke-White/Time Life Pictures/Getty Images: 66

Branger/Getty Images: 152a

William Thomas Cain/Getty Images: 104–105

Ed Clark/Time Life Pictures/Getty Images: 152b

Timothy A. Clary/AFP/Getty Images: 160

©CORBIS: 54, 57b, 68, 71, 78, 84

Cornell Formula SAE: 130a, 130b

Gordon Coster/Time Life Pictures/Getty Images: 90

Frank C. Duarte, Jr.: 97

Courtesy of Econogics: 33

James Ettaro, California State University Los Angeles, Department of Technology: 129

J. R. Eyerman/Time Life Pictures/Getty Images: 98, 101

Fox Photos/Getty Images: 26–27

FPG/Getty Images: 87

George Frey/Getty Images: 105

Rich Frishman/Time Life Pictures/Getty Images: 137

General Photographic Agency/Getty Images: 155

Courtesy, Georgia Archives: 35

Harold Lloyd Trust/Getty Images: 64–65

From the Collections of Henry Ford Museum & Greenfield Village & Ford Motor Company: 4, 40–41

HO/AFP/Getty Images: 79

Hulton Archive/Getty Images: 16–17, 19, 20, 28b, 29, 31, 34b, 38, 39b, 42, 43, 44, 47, 48, 51, 60, 61, 72–73, 80, 85, 156b

A. Jones/Express/Getty Images: 95c

John Kobal Foundation/Getty Images: 88–89

Dmitri Kessel/Time Life Pictures/Getty Images: 156a

Keystone/Getty Images: 93, 143a, 143b

Edwin Levick/Hulton Archive/Getty Images: 39a

Erik S. Lesser/Getty Images: 117

Tom G. Lynn/Time Life Pictures/Getty Images: 138b

Leonard McCombe/Time Life Pictures/Getty Images: 76a

©Minnesota Historical Society/CORBIS: 36

Ralph Morse/Time Life Pictures/Getty Images: 110b

Motoring Picture Library/Alamy: 17

MPI/Getty Images: 57a, 69, 70

Museum of the City of New York/Byron Collection/Getty Images: 30

©Museum of Flight/CORBIS: 75, 86

NASA: 96(KSC-69PC-0421)

NASA/Newsmakers/Getty Images: 157

New York Times Co./Getty Images: 94b

Robert Nickelsberg/Getty Images: 100

Courtesy of NJIT MINI-BAJA: 119

Paul Popper/Mansell/Time Life Pictures/Getty Images: 62–63

Mike Powell/Allsport/Getty Images: 139b

Private Collection: 65

Bill Pugliano/Getty Images: 106

Jose Luis Roca/AFP/Getty Images: 161

©Bill Ross/CORBIS: 12–13

©Charles E. Rotkin/CORBIS: 99

©Schenectady Museum; Hall of Electrical History Foundation/CORBIS: 153a, 153b

David E. Scherman/Time Life Pictures/Getty Images: 22

Tim Sloan/AFP/Getty Images: 67

Larry W. Smith/Getty Images: 146

Society of Automotive Engineers: 1, 2–3, 6, 24, 82a, 115, 120, 121, 131, 164a, 164b, 164c, 164d, 165a, 165b, 165c, 165d, 166a, 166b, 166c, 166d, 167a, 167b, 167c, 167d, 168a, 168b, 168c, 168d, 169a, 169b, 169c, 169d, 170a, 170b, 170c, 170d, 171a, 171b, 171c, 171d, 172a, 172b, 172c, 172d, 173a, 173b, 173c, 173d, 174a, 174b, 174c, 174d, 175a, 175b, 175c, 175d, 176a, 176b, 176c, 176d, 177a, 177b, 177c, 177d, 178a, 178b, 178c, 178d, 179a, 179b, 179c, 179d, 180a, 180b, 180c, 180d, 181a, 181b, 181c, 181d, 182a, 182b, 182c, 182d, 183a, 183b, 183c, 183d, 184A, 184b, 184c, 184d, 185a, 185b, 185c, 185d, 186a, 186b, 186c, 186d, 187a, 187b, 187c, 187d, 188a, 188b, 188c, 188d, 189a, 189b, 189c, 189d, 189e

©Joseph Sohm:Visions of America/CORBIS: 102–103

Mario Tama/Getty Images: 148–149

Time Magazine, Copyright Time Inc./Time Life Pictures/Getty Images: 142a

Topical Press Agency/Getty Images: 32, 58, 110a

Toyota/Getty Images: 107

©Underwood & Underwood/CORBIS: 59

The University of Akron: 126–127

U.S. Air Force/Getty Images: 159

U.S. Army Transportation Museum: 56, 74

Greg Wood/AFP/Getty Images: 150–151

Stefan Zaklin/Getty Images: 11